Recent Findings in Boolean Techniques

Rolf Drechsler • Daniel Große
Editors

Recent Findings in Boolean Techniques

Selected Papers from the 14th International Workshop on Boolean Problems

Editors
Rolf Drechsler
University of Bremen/DFKI
Bremen, Germany

Daniel Große
Johannes Kepler University of Linz
Linz, Austria

ISBN 978-3-030-68073-2 ISBN 978-3-030-68071-8 (eBook)
https://doi.org/10.1007/978-3-030-68071-8

This Springer imprint is published by the registered company Springer Nature Switzerland AG
The registered company address is: Gewerbestrasse 11, 6330 Cham, Switzerland

Preface

Boolean functions are at the core of computer science and the foundation of today's circuits and systems. The *International Workshop on Boolean Problems* (IWSBP) is a bi-annually held and a well-established forum to discuss the recent advances on problems related to Boolean logic and Boolean algebra. In 2020, the 14th edition of the workshop was held virtually from September 24 to September 25 due to the worldwide pandemic. The workshop provided a forum for researchers and engineers from different disciplines to exchange ideas as well as to discuss problems and solutions. The workshop is devoted to both theoretical discoveries and practical applications. This edited book contains a selection of best papers presented at the workshop and one additional paper. The papers in this volume demonstrate new accomplishments in the theory of Boolean problems. Furthermore, several papers illustrate how these results find their way into important practical applications.

The first two chapters in the book are contributions that resulted from the invited keynotes at the workshop. In Chap. 1, Daniela Kaufmann presents *Formal Verification of Integer Multiplier Circuits using Algebraic Reasoning—A Survey*. In Chap. 2, Victor M. van Santen, Florian Klemme, and Hussam Amrouch write about *The Vital Role of Machine Learning in Developing Emerging Technologies*. The following six chapters are extended manuscripts based on the workshop submissions. In Chap. 3, Bernd Steinbach and Christian Posthoff consider *Fast Optimal Synthesis of Symmetric Index Generation Functions*. Felix Weitkämper targets *Axiomatizing Boolean Differentiation* in Chap. 4. In Chap. 5, Radomir S. Stanković, Milena Stanković, Claudio Moraga, and Jaakko Astola investigate bent functions in *Construction of Binary Bent Functions by FFT-Like Permutation Algorithms*. In Chap. 6, Jan Schmidt and Petr Fišer write about *Nonlinear Codes for Test Patterns Compression: The Old School Way*. D. Michael Miller and Gerhard W. Dueck address *Translation Techniques for Reversible Circuit Synthesis with Positive and Negative Controls* in Chap. 7. In Chap. 8, Claudio Moraga focuses on *Hybrid Control of Toffoli and Peres Gates*. Finally, the book is concluded in Chap. 9 by Alireza Mahzoon, Daniel Große, and Rolf Drechsler with *GenMul: Generating Architecturally Complex Multipliers to Challenge Formal Verification Tools*.

We would like to express our thanks to the program committee of the 14th IWSBP as well as to the organizational team, in particular Alireza Mahzoon, Lisa Jungmann, and Kristiane Schmitt. Furthermore, we thank all the authors of contributed chapters who did a great job in submitting their manuscripts of very high quality. A special thanks goes to the keynote speakers of the workshop, Dr. Daniela Kaufmann (Johannes Kepler University Linz, Austria) and Junior Professor Dr. Hussam Amrouch (University of Stuttgart, Germany). Finally, we would like to thank Nandhakumar Sundar, Brian Halm, Zoe Kennedy, and Charles Glaser from Springer. All this would not have been possible without their steady support.

Bremen, Germany Rolf Drechsler

Linz, Austria Daniel Große
December 2020

Contents

Formal Verification of Integer Multiplier Circuits Using Algebraic Reasoning: A Survey 1
Daniela Kaufmann

The Vital Role of Machine Learning in Developing Emerging Technologies 29
Victor M. van Santen, Florian Klemme, and Hussam Amrouch

Fast Optimal Synthesis of Symmetric Index Generation Functions 59
Bernd Steinbach and Christian Posthoff

Axiomatizing Boolean Differentiation 83
Felix Weitkämper

Construction of Binary Bent Functions by FFT-Like Permutation Algorithms 105
Radomir S. Stanković, Milena Stanković, Claudio Moraga, and Jaakko Astola

Nonlinear Codes for Test Patterns Compression: The Old School Way ... 125
Jan Schmidt and Petr Fišer

Translation Techniques for Reversible Circuit Synthesis with Positive and Negative Controls 143
D. Michael Miller and Gerhard W. Dueck

Hybrid Control of Toffoli and Peres Gates 167
Claudio Moraga

GenMul: Generating Architecturally Complex Multipliers to Challenge Formal Verification Tools 177
Alireza Mahzoon, Daniel Große, and Rolf Drechsler

Index 193

Formal Verification of Integer Multiplier Circuits Using Algebraic Reasoning: A Survey

Daniela Kaufmann

1 Introduction

Digital circuits carry out logical operations, which make them an important component in computers and digital systems, because they represent models for various digital components and arithmetic operations. The basic function of a digital circuit is to compute binary digital values for the logical function it implements, given binary values at the input. The computation is usually realized by logic gates that represent simple Boolean functions, such as negation (NOT), conjunction (AND), disjunction (OR), or exclusive disjunction (XOR). These logic gates can be combined to build more complex logical operations. A subclass of digital circuits are combinational logic circuits, where the output of the circuit is a function of the present input only, i.e., the output does not depend on previous input values. Combinational logic is used in computer circuits to perform Boolean algebra. For example, the part of an arithmetic logic unit (ALU) in a CPU, which is responsible for mathematical calculations, is constructed using combinational logic. If a circuit implements an arithmetic operation, it is called an *arithmetic circuit*, which can be further refined to determine specific arithmetic operations such as *adder circuits* or *multiplier circuits*.

Since these circuits are such a crucial part of processors, it is extremely important to guarantee their correctness in order to prevent issues like the famous Pentium FDIV bug [49] that was detected in 1994. This bug affected the floating point unit of early Intel Pentium processors. The division algorithm for floating points used a lookup table to calculate the intermediate quotients. Due to a programming error, five entries of the lookup table contained zero instead of +2. Thus the result was

D. Kaufmann (✉)
Johannes Kepler University Linz, Linz, Austria
e-mail: daniela.kaufmann@jku.at

R. Drechsler, D. Große (eds.), *Recent Findings in Boolean Techniques*,
https://doi.org/10.1007/978-3-030-68071-8_1

incorrect and in the worst case the error could affect the fourth significant digit of a decimal number. Even more than 25 years after detecting this bug, automatically proving the correctness of arithmetic circuits, and especially multiplier circuits, is still considered to be a challenge.

Formal verification can be used to prove or disprove the correctness of a given system with respect to a predefined specification. To this end the system is translated into a mathematical model, and automated decision processes are applied to derive the desired correctness property. The different formal verification approaches are distinguished by the mathematical formalism used in the verification process.

Up to now several solving techniques have been developed for multiplier verification. The first technique that was shown to detect the Pentium bug is based on binary decision diagrams [10], more precisely on binary moment diagrams (BMDs) [13] and variants [14], since their size remains linear in the number of input bits of a multiplier. However, this approach requires structural knowledge of the multipliers [11, 13]. It is important to determine the order in which BMDs are built, because it has tremendous influence on the size and thus performance.

A common approach models the problem as a satisfiability (SAT) problem, where the circuit is translated into a formula in conjunctive normal form (CNF). A large set of such encodings was submitted to the SAT Competition 2016 [7]. The results indicated that verifying CNF miters of multipliers needs exponential-sized resolution proofs [8], which implies exponential run-time of CDCL SAT solvers. For simple multiplier architectures, this conjecture is neglected in theory in [5], where it was shown that ring properties do admit polynomial-sized resolution proofs. Recent work shows that pseudo-Boolean solvers can verify the word-level equivalence of simple multiplier architectures that consist only of half- and full-adders [32]. This method is so far not applicable for more complex architectures.

A further approach is based on the usage of theorem provers, such as ACL2 [26]. Theorem provers in combination with SAT are able to certify industrial multipliers [22]. Typically, theorem provers are not fully automated and require domain knowledge. Recently, progress has been made in the theorem prover ACL2 [52], which now allows automated verification of a large set of multiplier architectures. However, the multipliers have to be given as SVL netlists, which rely on the preservation of hierarchical information of the circuits.

Approaches based on bit-level reverse engineering [45, 50] use arithmetic bit-level representations, which are extracted from the gate-level netlists. They are able to verify simple multipliers, but fail to verify non-trivial multipliers. Methods based on term rewriting [53] require domain knowledge and thus are not fully automated.

The currently most effective technique for automated verification of flattened multipliers is based on computer algebra, e.g., [16, 24, 40]. In this method, all gates of the circuit and its specification are represented by polynomials. If the gate polynomials are ordered according to their topological appearance, they generate a Gröbner basis [12]. Hence, the question whether a multiplier circuit is correct can be answered by reducing the specification by the implied Gröbner basis. The multiplier is correct if and only if the reduction returns zero. The main issue of the general algebraic approach is that the size of the intermediate reduction results increases drastically. Thus, several preprocessing techniques and reduction methods have been developed in recent years [16, 24, 40], which attempt to overcome this issue.

Nonetheless, the verification process might not be error-free. Generating and checking proofs independently increases trust in the results of automated reasoning tools. Polynomial proofs can be obtained as a by-product of verifying multiplier circuits [25, 30] and can be checked by independent proof checking tools.

In this chapter, we survey over the current state of the art in verifying integer multipliers using computer algebra. For verification of Galois field multipliers, we refer to [33, 34, 58, 59]. In Sect. 2, we introduce the technique of circuit verification based on algebraic reasoning and present available proof formats. In Sect. 3, we present recent verification tools and discuss their strategies to overcome the issue of monomial blow-up in the intermediate reduction results. We show available benchmark generators in Sect. 4 and conclude with a comprehensive evaluation in Sect. 5.

2 Circuit Verification Using Computer Algebra

In this section, we introduce multiplier circuits and discuss architectural details. We present the algebraic concepts that are needed in the technique of automated circuit verification using computer algebra. Furthermore, we introduce algebraic proof systems that can be used to validate the correctness of the verification results.

2.1 Multiplier Circuits

A digital circuit implements a logical function and computes binary digital values, given binary values at the input. The computation of the function is realized by logic gates, such as NOT, AND, OR, and XOR. The specification of a circuit is the desired relation between its inputs and outputs. A circuit *fulfills a specification* if for all inputs it produces outputs that match this desired relation. The goal of verification is to formally prove that the circuit fulfills its specification.

In this chapter, we consider gate-level integer multipliers with input bits $a_0, \dots, a_{n-1}, b_0, \dots, b_{n-1} \in \{0, 1\}$ and $2n$ output bits $s_0, \dots, s_{2n-1} \in \{0, 1\}$. If the circuit represents multiplication of unsigned integers, the multiplier is correct if and only if for all possible inputs the specification $\mathcal{U}_n = 0$ holds, where:

$$\mathcal{U}_n = -\sum_{i=0}^{2n-1} 2^i s_i + \left(\sum_{i=0}^{n-1} 2^i a_i\right)\left(\sum_{i=0}^{n-1} 2^i b_i\right) \tag{1}$$

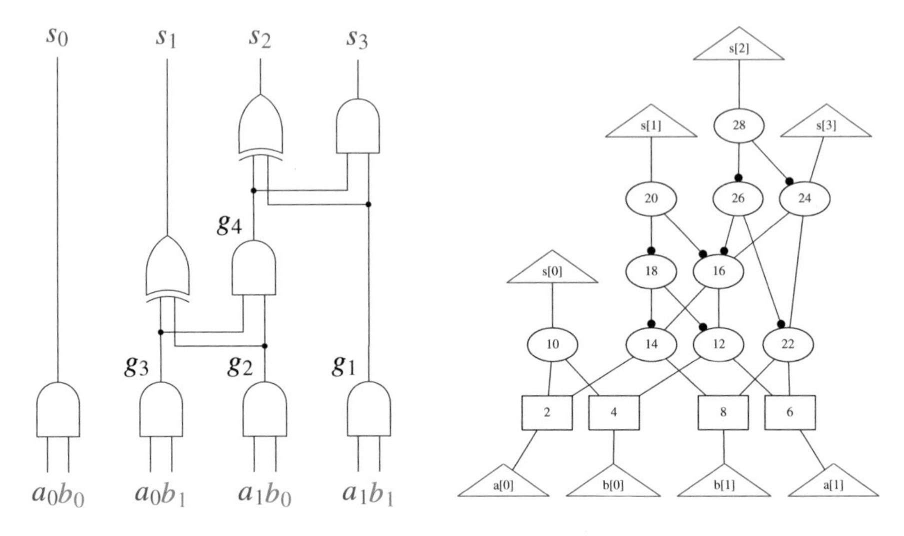

Fig. 1 Gate-level (left) and AIG (right) representation of a 2-bit multiplier circuit [24]

Example 1 The left side of Fig. 1 shows the gate-level representation of a 2-bit unsigned integer multiplier. The variables a_1, a_0, b_1, b_0 represent the input bits of the multiplier and s_3, s_2, s_1, s_0 are the binary outputs of the multiplier. The word-level specification of this circuit is $-8s_3 - 4s_2 - 2s_1 - s_0 + (2b_1 + b_0)(2a_1 + a_0) = 0$.

If the circuit represents signed multiplication, we have to take into account that the integers in the specification $\mathcal{S}_n$ are represented using two's complement.

$$\mathcal{S}_n = -2^{2n-1}s_{2n-1} + \sum_{i=0}^{2n-2} 2^i s_i - \left(-2^{n-1}a_{n-1} + \sum_{i=0}^{n-2} 2^i a_i\right)\left(-2^{n-1}b_{n-1} + \sum_{i=0}^{n-2} 2^i b_i\right) \tag{2}$$

A common representation of combinational circuits is the encoding as an and-inverter-graph (AIG) [31]. An AIG is a directed acyclic graph, which consists of two-input nodes representing logical conjunction. The edges may contain a marking that indicates logical negation. The AIG representation usually contains more nodes than the gate-level representation but has an unequivocal syntax and semantics and is very efficient to manipulate. The right side of Fig. 1 shows the AIG representation of the gate-level multiplier that is depicted on the left side.

The space and time complexity of a multiplier depends on its architecture. In general, a multiplier circuit can be divided into three parts [44]. In the first component, *partial product generation* (PPG), the partial products a_ib_j for $0 \leq i < n, 0 \leq j < n$, as contained in the specification, are generated. This can, for example, be achieved by using simple AND-gates or using a more complex Booth encoding [44].

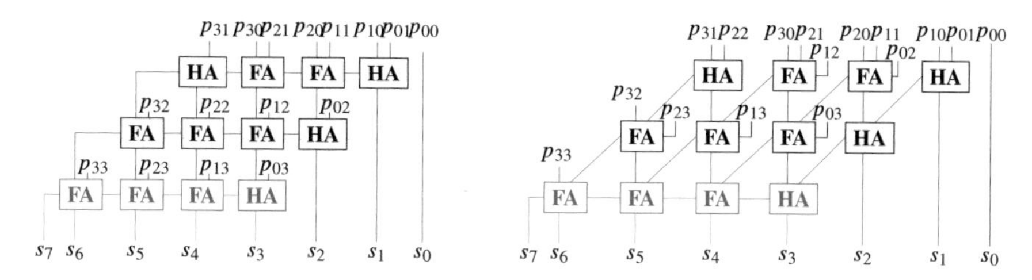

Fig. 2 Architecture of array multipliers (left) and diagonal multipliers (right) [24]

In the second component, *partial product accumulation* (PPA), the partial products are reduced to two layers by multi-operand addition using half-adders (HA), full-adders (FA), and compressors. Well-known accumulation structures are, for example, array or diagonal accumulation, Wallace trees, or compressor trees [44].

In the *final-stage adder* (FSA), the output of the circuit is computed using an adder circuit. Generally, adder circuits can be split into two groups: either the carries are computed alongside the sum bits or they are calculated before the sums. Adders of the first group consist of a sequence of half- and full-adders, giving them a simple but inefficient structure. Examples are ripple-carry or carry-select adders. In order to decrease the latency of carry computation, the adder circuits of the second group precompute the carry bits of the adder. They are called *generate-and-propagate (GP) adders*. Examples are carry look-ahead adders and Kogge-Stone adders [44].

We call multipliers, that can be fully decomposed into half- and full-adders *simple multipliers*, all other architectures are called *complex multipliers*.

Example 2 We show two simple multiplier architectures with input bit-width 4 in Fig. 2. In both circuits, the PPG uses AND-gates, i.e., $p_{ij} = a_i \wedge b_j$. In array multipliers, which are shown on the left side, the partial products are accumulated using a grid-like structure. The multiplier on the right side uses a diagonal structure. In both multipliers, the FSA is a ripple-carry adder, which is highlighted in red.

2.2 Algebra

Let us now briefly summarize algebraic concepts, following [18]. Throughout this section, let $\mathbb{K}[X] = \mathbb{K}[x_1, \ldots, x_n]$ denote the ring of polynomials in variables $x_1, \ldots, x_n$ with coefficients in a field $\mathbb{K}$.

Definition 1 A *term* τ is a product of the form $\tau = x_1^{e_1} \cdots x_n^{e_n}$ for $e_1, \ldots, e_n \in \mathbb{N}$. A *monomial* $m = \alpha\tau$ is a constant multiple of a term, with $\alpha \in \mathbb{K}$. A *polynomial* $p = m_1 + \cdots + m_s$ is a finite sum of monomials.

On the set of terms, an order $\leq$ is fixed such that for all terms τ, σ_1, σ_2 we have $1 \leq \tau$ and $\sigma_1 \leq \sigma_2 \Rightarrow \tau\sigma_1 \leq \tau\sigma_2$. A term order is called a *lexicographic term order* if for all terms $\sigma_1 = x_1^{u_1} \cdots x_n^{u_n}, \sigma_2 = x_1^{v_1} \cdots x_n^{v_n}$ we have $\sigma_1 < \sigma_2$ if and only if there exists an index i with $u_j = v_j$ for all $j < i$, and $u_i < v_i$. Every polynomial $p \neq 0$ contains only finitely many terms, the largest of which (with respect to the chosen order $\leq$) is called the *leading term* and denoted by $\operatorname{lt}(p)$. If $p = \alpha\tau + \cdots$ and $\operatorname{lt}(p) = \tau$, then $\operatorname{lc}(p) = \alpha$ is called the *leading coefficient* and $\operatorname{lm}(p) = \alpha\tau$ is called the *leading monomial* of p. We call $p - \alpha\tau$ the *tail* of p.

Definition 2 A nonempty set $I \subseteq \mathbb{K}[X]$ is called an *ideal* if $\forall\, p, q \in I : p + q \in I$ and $\forall\, p \in \mathbb{K}[X]\ \forall\, q \in I : pq \in I$. If $I \subseteq \mathbb{K}[X]$ is an ideal, then a set $P = \{p_1, \ldots, p_m\} \subseteq \mathbb{K}[X]$ is called a *basis* of I if $I = \{q_1 p_1 + \cdots + q_m p_m \mid q_1, \ldots, q_m \in \mathbb{K}[X]\}$. We say *I is generated by P* and write $I = \langle P \rangle$.

The theory of Gröbner bases offers a decision procedure for the so-called ideal membership problem, i.e., given $q \in \mathbb{K}[X]$ and a basis $P = \{p_1, \ldots, p_m\} \subseteq \mathbb{K}[X]$, decide whether q belongs to the ideal generated by $p_1, \ldots, p_m$. If $\{p_1, \ldots, p_m\}$ is a Gröbner basis, then the question can be answered using a multivariate version of polynomial division with remainder (cf. Thm. 3 in Chap. 2 §3 of [18]).

Definition 3 A basis $P = \{p_1, \ldots, p_m\}$ of an ideal $I \subseteq \mathbb{K}[X]$ is called a *Gröbner basis* (with respect to a fixed order $\leq$) if and only if $\forall q \in I \exists p_i \in P : \operatorname{lm}(p_i) \mid \operatorname{lm}(q)$.

Lemma 1 *Every ideal $I \subseteq \mathbb{K}[X]$ has a Gröbner basis with respect to a fixed order $\leq$.*

Proof Cor. 6 in Chap. 2 §5 of [18]. □

Given an arbitrary basis of an ideal, Buchberger's algorithm [12] is able to compute a Gröbner basis for it in finitely many steps.

Lemma 2 *If $P = \{p_1, \ldots, p_m\}$ is a Gröbner basis, then every $f \in \mathbb{K}[X]$ has a unique remainder r with respect to P. Furthermore, it holds that $f - r \in \langle P \rangle$.*

Proof Prop. 1 in Chap. 2 §6 of [18]. □

Ultimately the following Lemma provides the answer on how we can solve the ideal membership problem with the help of Gröbner basis and thus can check whether a polynomial belongs to an ideal or not.

Lemma 3 *Let $P = \{p_1, \ldots, p_m\} \subseteq \mathbb{K}[X]$ be a Gröbner basis, and let $f \in \mathbb{K}[X]$. Then f is contained in the ideal $I = \langle P \rangle$ if and only if the remainder of f with respect to P is zero.*

Proof Cor. 2 in Chap. 2 §6 of [18]. □

2.3 Circuit Verification Using Computer Algebra

In this section we introduce the technique of circuit verification using computer algebra, following [27]. We consider circuits C with inputs $a_0, \ldots, a_{n-1}$ and $b_0, \ldots, b_{n-1}$, outputs $s_0, \ldots, s_{2n-1}$, and a number of logical gates, denoted by $g_1, \ldots, g_k$. By R we denote the ring $\mathbb{K}[a_0, \ldots, a_{n-1}, b_0, \ldots, b_{n-1}, g_1, \ldots, g_k, s_0, \ldots, s_{2n-1}] = \mathbb{K}[X]$.

The semantics of each circuit gate implies a polynomial relation among the input and output variables, such as the following ones:

$$\begin{array}{lll} u = \neg v & \text{implies} & 0 = -u + 1 - v \\ u = v \wedge w & \text{implies} & 0 = -u + vw \\ u = v \vee w & \text{implies} & 0 = -u + v + w - vw \\ u = v \oplus w & \text{implies} & 0 = -u + v + w - 2vw. \end{array} \tag{3}$$

We call these polynomials *gate polynomials* or *gate constraints*. Let $G(C) \subseteq R$ denote the set of polynomials, which contains for each gate of the given circuit the corresponding polynomial of (3).

Example 3 The possible solutions for the gate constraint $p_{00} = a_0 \wedge b_0$ represented as (p_{00}, a_0, b_0) are $(1, 1, 1), (0, 1, 0), (0, 0, 1), (0, 0, 0)$ which are all solutions of the polynomial $-p_{00} + a_0 b_0 = 0$, when a_0, b_0 are restricted to the Boolean domain.

All variables $x \in X$ are Boolean and we enforce this property by assuming the set $B(X) = \{x(1-x) \mid x \in X\} \subseteq R$ of *Boolean value constraints*.

Since the logical gates are functional, the values of $g_1, \ldots, g_k, s_0, \ldots, s_{2n-1}$ in a circuit are determined as soon as the inputs $a_0, \ldots, a_{n-1}, b_0, \ldots, b_{n-1} \in \{0, 1\}$ are fixed. This motivates the following definition of polynomial circuit constraints [27].

Definition 4 Let C be a circuit. A polynomial $p \in R$ is called a *polynomial circuit constraint (PCC)* for C if for every choice of

$$(a_0, \ldots, a_{n-1}, b_0, \ldots, b_{n-1}) \in \{0, 1\}^{2n}$$

and the resulting values $g_1, \ldots, g_k, s_0, \ldots, s_{2n-1}$ which are implied by the gates of the circuit C, the substitution of all these values into the polynomial p gives zero. The set of all PCCs for C is denoted by $I(C)$.

It is easy to see that $I(C)$ is an ideal of R. Since it contains all PCCs, this ideal includes all relations that hold among the values at the different points in the circuit. The circuit fulfills a certain specification $\mathcal{L}$ if and only if the polynomial relation corresponding to the specification of the circuit is contained in the ideal $I(C)$.

Thus, checking whether a given circuit C is a correct multiplier reduces to an ideal membership test. Definition 4 does not provide any information of a basis of $I(C)$, hence Gröbner basis technology is not directly applicable. However, we can deduce at least some elements of $I(C)$ from the semantics of circuit gates.

Definition 5 Let C be a circuit and assume $G(C) \subseteq R$ be the set which contains for each gate of C the corresponding polynomial of (3).

Let $B_0(G) = B(\{a_0, \ldots, a_{n-1}, b_0, \ldots, b_{n-1}\})$ and $J(C) = \langle G(C) \cup B_0(G) \rangle \subset R$.

Assume that we have a verifier which checks for a given circuit C and a given specification polynomial $\mathcal{L} \in R$ whether it holds that $\mathcal{L} \in J(C)$. Because it holds that $J(C) = I(C)$ [27], such a verifier is sound and complete.

Theorem 1 *Let C be a circuit, and let $J(C)$ be as in Definition 5. Furthermore, let $\leq$ be a reverse topological lexicographic term order where the variables are ordered such that the variable of a gate output is always greater than the variables attached to the input edges of that gate. Then $G(C) \cup B_0(G)$ is a Gröbner basis for $J(C)$ with respect to the ordering $\leq$.*

Proof This theorem is shown for instance in [27, 34, 54].

Hence $G(C) \cup B_0(G)$ is a Gröbner basis for the ideal $J(C)$ and we can decide membership using Gröbner bases theory, i.e., we reduce the specification $\mathcal{L}$ by elements of $G(C) \cup B_0(G)$ until no further reduction is possible. The circuit is correct if and only if the final remainder is zero.

In this section we restricted the theory to polynomial rings over a field $\mathbb{K}$. A generalization for polynomial rings over *principal ideal domains* (such as $\mathbb{Z}$) can be found in [28], where it is furthermore discussed how to invoke modular reasoning, i.e., reasoning in rings $R = \mathbb{Z}_l[X]$. Modular reasoning allows to eliminate monomials that have large coefficients.

As a final remark, in the case when a polynomial g is not contained in an ideal $I = \langle P \rangle$, i.e., the remainder of dividing g by P is not zero and allows to determine a concrete choice of input assignments for which g does not vanish. In our application of multiplier verification, these evaluations provide counter-examples, in case a circuit is determined not to be a multiplier.

2.4 Algebraic Proof Systems

Although the verification method is sound and complete, it may happen that the implementation contains errors and the reasoning engine delivers wrong results. One way to overcome this issue is to verify the implementation, e.g., [52], which is typically very tedious and requires a lot of effort. Thus, it is common to produce proof certificates in the reasoning engine to monitor the verification process. These proofs are generated as by-product of the reasoning technique and are given to independent (and ideally verified) proof checkers to validate the verification result.

For computer algebra, two algebraic proof systems are used in practice, the *practical algebraic calculus* (PAC) [46], which is based on the *polynomial calculus* (PC) [17], and the *Nullstellensatz proof* format (NSS) [4].

Practical Algebraic Calculus The practical algebraic calculus [46] is an instantiated version of PC [17] which allows efficient proof checking. A PAC proof consists of three components (i) the given set of polynomials G, i.e., the constraint set, (ii) the core proof, i.e., a sequence of proof rules P that model the properties of an ideal, and (iii) the target polynomial f. In a correct proof, it is derived whether the target polynomial can be derived from the constraint set using the proof rules.

Initially the proof format has been defined for polynomial rings $\mathbb{K}[X]$, where $\mathbb{K}$ is a field [17, 46] and all variables represent Boolean values.

The soundness and completeness arguments have been generalized to rings $R[X]$, where all polynomials in the constraint set have unique leading terms that contain only a single variable, cf. Thm. 1 and Thm. 2 in [28]. Recently, the PAC format has been revised to derive a more compact proof representation [30].

Let P be a sequence of polynomials that can be accessed via indices. We write $P(i) = \bot$ to denote that the sequence P at index i does not contain a polynomial, and $P(i \mapsto p)$ to determine that P at index i is set to p. The initial state is $(X = \text{Var}(G \cup \{f\}), P)$ where P maps indices to polynomials of G.

[ADD (i, j, k, p)] $\quad (X, P) \implies (X, P(i \mapsto p))$

where $P(j) \neq \bot$, $P(k) \neq \bot$, $P(i) = \bot$, $p \in R[X]/\langle B(X)\rangle$, and $p = P(j) + P(k)$.

[MULT (i, j, q, p)] $\quad (X, P) \implies (X, P(i \mapsto p))$

where $P(j) \neq \bot$, $P(i) = \bot$, $p, q \in R[X]/\langle B(X)\rangle$, and $p = q \cdot P(j)$.

PAC proofs that are defined over $\mathbb{Z}[X]$ can be checked by the checkers PACHECK (implemented in C) [30] and PASTÈQUE (verified in Isabelle/HOL) [30].

Nullstellensatz The Nullstellensatz proof system [4] allows to derive whether a target polynomial $f \in R[X]$ can be represented as a linear combination from a given set of polynomials $G = \{g_1, \ldots, g_l\} \subseteq R[X]$ and the Boolean value constraints $B(X)$. Similar to PAC, the NSS proof system is initially defined for polynomial rings over fields [4]. By the same arguments given for PAC, the soundness and completeness arguments can be generalized for rings $R[X]$ where all polynomials in G have unique leading terms that contain only one variable [25].

Again, we handle the Boolean value constraints implicitly and derive the following proof format. For a polynomial $f \in R[X]/\langle B(X)\rangle$ and a given set of polynomials $G = \{g_1, \ldots, g_l\} \subseteq R[X]/\langle B(X)\rangle$, an NSS proof is an equality P, such that

$$\sum_{i=1}^{l} h_i g_i = f \in R[X]/\langle B(X)\rangle, \tag{4}$$

with $h_i \in R[X]/\langle B(X)\rangle$.

Nullstellensatz proofs over $\mathbb{Z}[X]$ can be checked using NUSS-CHECKER [25].

3 Verification Tools

In this section we present reasoning tools for verification of flattened gate-level integer multipliers using computer algebra. We focus on the most recent work that has been developed in the last 3 years and consider the tools from Yu et al.: ABC/ARTI [16, 60]; Kaufmann et al.: AMULET [28]; and Mahzoon et al.: POLYCLEANER [38], REVSCA/REVSCA-2.0 [40], DYPOSUB [42] (sorted chronologically).

We discuss the scope of application of these tools and present their techniques that help to overcome the issue of monomial blow-up during reduction. All these tools are considered in the experimental evaluation in Sect. 5.

3.1 Algebraic RewriTing in ABC [15, 16, 57, 60]

The authors of [15, 57] use a method called *function extraction* to verify circuits. Function extraction is a similar algebraic approach to Gröbner basis reduction as presented in Sect. 2. The difference to Gröbner basis reduction is that it is not required to provide the complete specification polynomial of the circuit for reduction. Instead the word-level output of the circuit, i.e., the bit-vector $\sum_{i=0}^{2n-1} 2^i s_i$ for unsigned numbers resp. $-2^{2n-1}s_{2n-1} + \sum_{i=0}^{2n-2} 2^i s_i$ for signed number representation, is reduced by the gate constraints of the given circuit. The Boolean value constraints are reduced implicitly, i.e., every exponent greater than one is immediately reduced to one. This method returns a unique polynomial representation of the functionality of the circuit in terms of the circuit inputs. In order to verify correctness of a circuit, this remainder polynomial needs to be compared to the desired circuit functionality.

In follow-up works [16, 60], the authors introduced an optimization, where half- and full-adders are extracted by identifying subcircuits in the given circuit that represent MAJ3 and XOR3 gates. These XOR3 and MAJ3 gates are essential components of adder trees that are present in most arithmetic circuits. The polynomial constraints of all circuit gates that belong to a MAJ3 or XOR3 gate are replaced by a single polynomial that encodes a MAJ3 or XOR3 gate in order to simplify the polynomial representation of the circuit. These polynomials are sorted topologically pairwise.

The authors developed a framework called ARTI (Algebraic RewriTing) [56] that is integrated within the ABC tool [6]. Verification of multipliers that are given as AIGs is executed using the command `&polyn`, which can be configured to define whether signed or unsigned multiplication is considered. The extraction of the adder trees is invoked by the command `&atree` [16, 60].

This technique is able to handle very large multipliers that can be fully decomposed into half- and full-adders (for instance, the array and diagonal multipliers of Fig. 2) efficiently, but fails on slightly optimized multiplier architectures, because invoking the command `&atree` on these multipliers leads to incompleteness. In the experimental evaluation, we use `&atree` only for those benchmarks where we know that the circuit can be represented by adder trees, i.e., the experiments of Sect. 5.5.

3.2 AMULET [28]

In [28], it is presented how the verification approach can be generalized to polynomial rings that include modular reasoning, i.e., $\mathbb{Z}_l[X]$ for $l = 2^{2n}$. This allows to cancel monomials in the intermediate results with coefficients that are multiples of 2^{2n}.

The technique of [28] uses an incremental verification algorithm for circuit verification. In this method, the multiplier circuit is divided into column-wise slices and the specification polynomial is split into multiple polynomials. The correctness of the circuit is shown by incrementally verifying the correctness of each slice. The main advantage of this approach is that only one small part of the global specification is used for reduction, which helps to reduce the size of the intermediate results.

Furthermore, variable elimination is applied before reduction, i.e., after assigning the gates to slices, all variables that occur in only one other polynomial within the same slice are eliminated. Structures that implement a Booth encoding are detected by pattern matching, and their internal gates are eliminated too.

After variable elimination, the column-wise specifications are reduced by the rewritten Gröbner basis until completion. However, certain parts of the multiplier, more precisely particular final stage adders, are hard to verify using computer algebra. These adders usually contain sequences of OR-gates, which lead to an exponential blow-up of the intermediate reduction results. On the other hand, equivalence checking of adders is easy for SAT [29].

We will take a quick excursion and introduce the SAT problem following [19]:

- A *literal* l is either a positive Boolean variable x or its negation $\overline{x}$.
- A *clause* C is a finite disjunction of literals.
- A *formula in conjunctive normal form* (CNF) F is a finite conjunction of clauses.
- An *assignment* τ is a function that consistently maps the literals of F to $v \in \{\mathbf{t}, \mathbf{f}\}$, such that $\tau(x) = v \Leftrightarrow \tau(\overline{x}) = \neg v$, where $\neg\mathbf{t} = \mathbf{f}$ and $\neg\mathbf{f} = \mathbf{t}$.
- A formula evaluates to **t** if and only if every clause in the formula evaluates to **t**. A clause C evaluates to **t** if $\exists l \in C$ with $\tau(l) = \mathbf{t}$. Given a CNF formula F, the SAT problem is to decide if there exists an assignment such that F evaluates to **t**. If such an assignment exists, the formula is *satisfiable*, otherwise it is *unsatisfiable*.

Based on the observation, the technique of [28] combines SAT and computer algebra. It is detected whether a multiplier contains a complex final stage adder, which is then replaced by a simple ripple-carry adder. A bit-level miter, which is expected to be unsatisfiable, is produced to verify the correctness of the replacement using SAT solvers, and the rewritten multiplier is verified by computer algebra techniques.

The authors implemented a tool called AMULET [28] that is able to handle signed and unsigned multipliers given as AIGs. The tool automatically applies adder substitution and verifies the (rewritten) multiplier using computer algebra. Furthermore, AMULET is so far the only tool that is able to produce proof certificates in the PAC and NSS proof formats, cf. Sect. 2.4. In the experiments of this chapter, we will use the maintained version AMULET 1.5 that is currently available on GitHub [23].

3.3 POLYCLEANER [38]

The work of [38] provides an extensive analysis why the number of monomials in the reduction results increases drastically, when no preprocessing is applied. The reason of the size explosion are certain monomials, called vanishing monomials, that reduce to zero later in the reduction process. The authors observe that the vanishing monomials origin from gates where the sum and carry output of a half-adder converge and the vanishing monomials remain in the intermediate reduction results until all internal gate polynomials of the half-adders have been substituted.

The work of [38] proposes a method where the converging gates are identified, and the vanishing monomials are locally removed before the specification is reduced. First, for each occurring half-adder in the circuit, possible converging gates are identified and the belonging input cones are determined. A polynomial is extracted for each converging gate by substituting the gate polynomials of the associated cone. Since the extracted polynomial contains the product of the sum and carry output of the corresponding half-adder, it contains the vanishing monomials. After locally removing these vanishing monomials, the specification polynomial is reduced by these vanishing-free polynomials to verify the circuit.

This method is implemented in the tool POLYCLEANER [37] that is able to verify unsigned multipliers that are given as flattened Verilog modules.

3.4 REVSCA/REVSCA-2.0 [40]

The paper [40] is a follow-up work on [38]. The authors elaborate on the disadvantages of the proposed method of [38]. First, the method of [38] highly depends on the detection of the half-adders that are considered to be implemented as pairs of XOR and AND gates. The second disadvantage is that the search space for finding the converging gates is very large. Consequently, the method of [38] only works for those multipliers where all half-adders can be detected and fails for more complex multiplier circuits.

In [40] the authors generalize the detection of converging gates and propose a technique that identifies so-called atomic blocks of the multiplier, i.e., half-adders, full-adders, and compressors, using reverse engineering. The detection of converging gates becomes more independent of the actual design of the atomic blocks and the search space to identify these gates can be limited. Furthermore, since not only half-adders are considered as atomic blocks, vanishing-free polynomials are not only generated for converging gate cones, but also for the outputs of atomic blocks and lead to a more compact polynomial representation.

The authors implemented a tool called REVSCA [39] that verifies unsigned multipliers given as AIGs. The implementation has been improved and additionally verification of signed multipliers is supported in REVSCA-2.0 [39].

3.5 DYPOSUB *[42]*

In contrast to the already described methods, the technique of [42], a follow-up work of [40], explicitly tries to tackle the problem of verifying multiplier circuits, where logic synthesis and technology mapping are applied.

The problem with these optimized multipliers is that the clear boundaries of certain substructures, such as internal half- and full-adders, may be blurred. Consequently, the compact representation of these internal substructures are no longer available. For example, the discussed methods of Sects. 3.1 and 3.2 heavily rely on these boundaries, either during rewriting or defining the substitution order.

The method described in [42] tries to overcome this issue by using a *dynamic substitution order* that allows to keep the size of the intermediate reduction results on a moderate level. Before reduction is applied, the circuit gates are preprocessed as described in [40], where atomic blocks and converging gate cones are identified and vanishing-free polynomials are extracted for these cones and atomic blocks.

After preprocessing, a dynamic backward rewriting approach is applied to verify the circuit. The core idea is that for each substitution step, a number of candidate polynomials is available, which maintains the overall topological sorting and guarantees that the polynomials of atomic blocks are substituted consecutively. This ensures that the gate polynomials only need to be considered once during reduction.

At each backward rewriting step, there may be several such possible candidates available. The dynamic backward rewriting chooses the reduction candidate by the number of the occurrences of the leading term in the intermediate reduction result in ascending order. After each substitution step, the increase in the number of monomials is checked. If the number of monomials grows by more than 10%, the step is undone and the specification is reduced by the next candidate polynomial in line. If there is no reduction of any candidate satisfying the threshold limit, the threshold limit is increased and the process is repeated from the first candidate.

This approach is implemented in the tool DYPOSUB [41] that verifies unsigned multipliers given as AIGs. The experimental data of [42] shows that this technique is able to verify industrial benchmarks created by Synopsis. Unfortunately, these benchmarks are not open source and are not considered in Sect. 5.

4 Benchmark Generators

The evaluation of tools heavily relies on reproducible experiments and thus on publicly available benchmarks that allow to evaluate and compare the performance of the verification tools. In this section, we present the currently most common used benchmarks suites, discuss how these benchmarks are processed to AIGs (cf. Sect. 4.6), and how they can be optimized using technology mapping in Sect. 4.7. An experimental evaluation of these benchmarks will be given in Sect. 5.

4.1 Arithmetic Module Generator

To the best of our knowledge, the web-based tool Arithmetic Module Generator [20, 21] offers the most comprehensive benchmark suite. The following components can be combined to gain 192 different multipliers, given as Verilog modules:

Part. Product Gen.	Part. Product Accum.	Final-Stage adder
Booth radix-4	Array	Block carry look-ahead
Simple (AND gates)	Balanced delay tree	Brent-Kung
	Dadda tree	Carry look-ahead
	Overturned-stairs tree	Carry select
	Red. binary addition tree	Carry-skip fix size
	Wallace tree	Carry-skip variable size
	(4;2) compressor tree	Conditional sum
	(7,3) counter tree	Han-Carlson
		Kogge-Stone
		Ladner-Fischer
		Ripple carry
		Ripple-block carry look-ahead

All architectures can be generated for unsigned and signed number representation, thus yielding a total of 384 multiplier architectures. The multipliers can be generated up to input bit-width 64. In addition to integer multipliers, the Arithmetic Module Generator also allows the generation of two- and multioperand adders and multiply accumulators as well as Mastrovito and Massey-Omura parallel multipliers.

4.2 GenMul

The tool GenMul is either available as a web-based [35] or as a stand-alone tool [36] and allows to generate unsigned and signed multiplier circuits in Verilog. The following components can be combined to gain 24 architectures:

Part. Product Gen.	Part. Product Accum.	Final-Stage adder
Simple (AND gates)	Array	Brent-Kung
	Counter-based Wallace tree	Carry look-ahead
	Dadda tree	Carry-skip
	Wallace tree	Kogge-Stone
		Ladner-Fischer
		Ripple carry

For signed multiplication, only 23 architectures are available, as the combination "Simple–Array–Carry-skip" yields an error message. The input bit-width can be set arbitrarily large; However, for input bit-widths that are larger than 128, the generation process may run for several minutes.

4.3 MultGen

MultGen is a stand-alone tool [51] and the following components can be combined to gain 24 architectures in Verilog:

Part. Product Gen.	Part. Product Accum.	Final-Stage adder
Booth radix-2	Dadda tree	Han-Carlson
Booth radix-4	Wallace tree	Kogge-Stone
Simple (AND gates)		Ladner-Fischer
		Ripple carry

All multipliers can be generated for signed and unsigned multiplication and the input bit-width can be selected arbitrarily large. In addition, MultGen is not only able to generate stand-alone multipliers, but is also able to merge four smaller stand-alone multipliers to receive a bigger multiplier. Furthermore, MultGen provides access to a fused multiply-add operation and to a generator for the dot product.

4.4 EPFL Combinational Benchmark Suite

The EPFL Combinational Benchmark Suite [1–3] has been introduced with the aim of defining a comparative standard for logic optimization and synthesis. The benchmark suite encomparates 23 combinational circuits and is divided into arithmetic, random/control, and MtM ("More than ten Million gates") parts. The arithmetic benchmarks cover a variety of arithmetic operations, such as square-root computation, division, and multiplication. The arithmetic benchmarks come in different bit-widths to provide diversity in the implementation complexity; the contained multiplier circuit has an input bit-width of 64. Each circuit is distributed in Verilog, VHDL, BLIF, and AIG formats.

4.5 ABC/BOOLECTOR

Both tools ABC [6] and BOOLECTOR [43] provide access to a simple array multiplier that leads to structural equivalent AIGs. The only difference is that the inputs are sorted in sequence in ABC, whereas they are sorted interleaved in BOOLECTOR.

A multiplier is generated in ABC using the command `gen` with option `-m`, which receives the input bit-width n as parameter `-N`. The command `gen` furthermore allows the generation of a ripple-carry adder, sorter, mesh, or a random single-output function. The BLIF format is converted into an AIG by structural hashing (using the command `strash` of ABC).

```
abc -c "gen -N $n -m abc${n}.blif"
abc -c "read abc${n}.blif" -c strash -c "write
                                        abc${n}.aig"
```

If we generate multipliers using the SMT-solver BOOLECTOR, we have to provide an SMT encoding of the multiplier circuit (with input bit-width n). The SMT encoding is then processed by BOOLECTOR to generate the AIG circuit:

```
m=`expr 2 \* $n`
btor=btor${n}.btor
echo "1 var $n a" >> $btor
echo "2 var $n b" >> $btor
echo "3 uext $m 1 $n" >> $btor
echo "4 uext $m 2 $n" >> $btor
echo "5 mul $m 3 4" >> $btor
id=6
i=0
while [ $i -lt $m ]
do
  slice=$id
  echo "$id slice 1 5 $i $i" >> $btor
```

```
    id='expr $id + 1'
    echo "$id root 1 $slice" >> $btor
    id='expr $id + 2'
    i='expr $i + 1'
  done

  boolector $btor -rwl=1 -dai > btor${n}.aig
```

4.6 *Processing Verilog Benchmarks*

The available benchmark generators MULTGEN, GENMUL, and the ARITHMETIC MODULE GENERATOR output circuits in Verilog. Most of the verification tools rely on an AIG as input format, because AIGs are easy to handle and have an unequivocal syntax and semantics. Thus, after generating the multipliers, the Verilog format needs to be processed to gain the AIG representation. We use the YOSYS OPEN SYNTHESIS SUITE [55] to convert the circuits from Verilog to BLIF. The command, which can be seen below, has been thankfully explained to us by Mathias Soeken.

First, the design hierarchy is checked, which is then flattened and a generic technology mapper is applied, to replace cells in the design. The output is printed in the BLIF format, which is translated to an AIG using structural hashing in ABC:

```
yosys -p "hierarchy -auto-top -check"
     -p flatten -p techmap -o <output.blif> <input.v>

abc -c "read <output.blif>" -c strash -c "write
                                   <output.aig>"
```

4.7 *Optimizing Benchmarks*

The multipliers that are generated using the tools presented in Sects. 4.1–4.5 follow a general structure, where the boundaries of internal structures, such as half- and full-adders, can be identified. In industrial benchmarks, these circuits are optimized to reduce the number of gates and to reduce the delay of the circuits. As already mentioned in Sect. 3.5, these benchmarks are often not publicly available.

As a compromise, we can optimize the generated circuits, e.g., in ABC, and apply rewriting and technology mapping. Optimizing the benchmarks has been thankfully explained to us by Maciej Ciesielski.

Rewriting without technology mapping is applied as follows:

```
abc -c "read <input.aig>"
    -c <syn>
    -c "write <output.aig>"
```

where <syn> is for instance replaced by dc2, resyn, resyn2, or resyn3, or even a combination of them.

The option dc2 applies combinational AIG optimization. The options resyn, resyn2, and resyn3 are standard scripts that are included in the file “abc.rc” of the source code of ABC. In these scripts, multiple rounds of technology-independent rewriting, refactoring, and restructuring of the AIG are performed.

If technology mapping should be involved, a standard cell library needs to be provided, e.g., "mcnc.genlib" of SIS [48] that is available at [47]. The cell library is parsed, and technology mapping can be applied by the command map. Structural hashing is applied before the output is printed:

```
abc -c "read <input.aig>"
    -c <syn>
    -c "read <genlib>"
    -c "map"
    -c strash
    -c "write <output.aig>"
```

5 Evaluation

In this chapter, we evaluate the presented tools ABC, AMULET 1.5, DYPOSUB, POLYCLEANER, REVSCA/REVSCA-2.0 of Sect. 3 using the benchmarks we have presented in Sect. 4. In our experiments, we use an Intel Xeon E5-2620 v4 CPU at 2.10 GHz (turbo-mode disabled) with a memory limit of 128 GB.

We measure the time from starting a tool until it is finished and the time is listed in rounded seconds (wall-clock time). For a run of AMULET 1.5, where several tool applications are used in the verification flow, we measure the time AMULET needs to apply adder substitution and circuit verification and include the time the SAT solver KISSAT [9] needs to verify the equivalence of the adders.

5.1 ARITHMETIC MODULE GENERATOR

In our first experiment, we consider the 384 unsigned and signed 64-bit multipliers that are generated using ARITHMETIC MODULE GENERATOR [20]. We process the Verilog format to AIG as discussed in Sect. 4.7. The tool POLYCLEANER reads flattened Verilog modules. We flatten the circuit in YOSYS, using the presented command in Sect. 4.7. We write the design to a Verilog file by exchanging the suffix “.blif” to “.v”, i.e., we write the circuit to <output.v>.

We set the time limit in the experiments to 300 s. The results can be seen in the CDF-plots that are shown in Fig. 3.

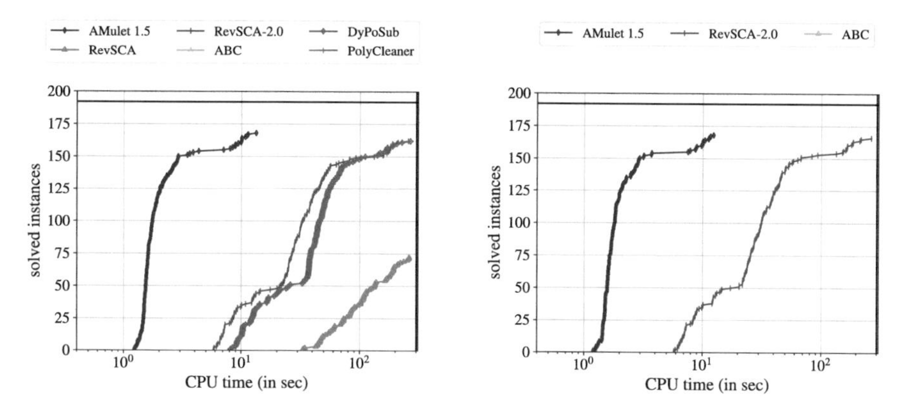

Fig. 3 Verification time (in sec) of unsigned (left) and signed (right) multipliers that are generated using the ARITHMETIC MODULE GENERATOR. Time limit: 300 s

All presented tools can be applied for verification of unsigned multipliers. However, only ABC, AMULET 1.5, and REVSCA-2.0 support verification of signed multipliers. These are the only tools used in the right part of Fig. 3.

It can be seen that AMULET 1.5, DYPOSUB, and REVSCA-2.0 solve the most benchmarks. More precisely, AMULET 1.5 solves 170 unsigned instances, DYPOSUB and REVSCA-2.0 solve 165 unsigned benchmarks. In both experiments, AMULET 1.5 is the fastest tool by an order of magnitude.

As discussed in Sect. 3.1, we do not apply the optimization `&atree`, which has the effect that ABC produces an internal error or a segmentation fault for all benchmarks. Thus there are no results for ABC. The tool POLYCLEANER exceeds the time limit in all experiments in the left side of Fig. 3.

5.2 GENMUL

In our second experiment, we consider the 24 unsigned and 23 signed benchmarks of GENMUL. We generated the benchmarks for an input bit-width of 64 and set the time limit to 300 s. The results are depicted in Fig. 4.

In the left side of Fig. 4, it can be seen that DYPOSUB and REVSCA-2.0 both solve 21 benchmarks, with REVSCA-2.0 being slightly faster. AMULET 1.5 solves 20 instances. ABC is an order of magnitude faster, but is only able to solve two instances. Again, POLYCLEANER exceeds the time limit in all experiments.

For the signed multipliers, ABC is again an order of magnitude faster, but only verifies 6 benchmarks. AMULET 1.5 and REVSCA-2.0 both solve 20 instances.

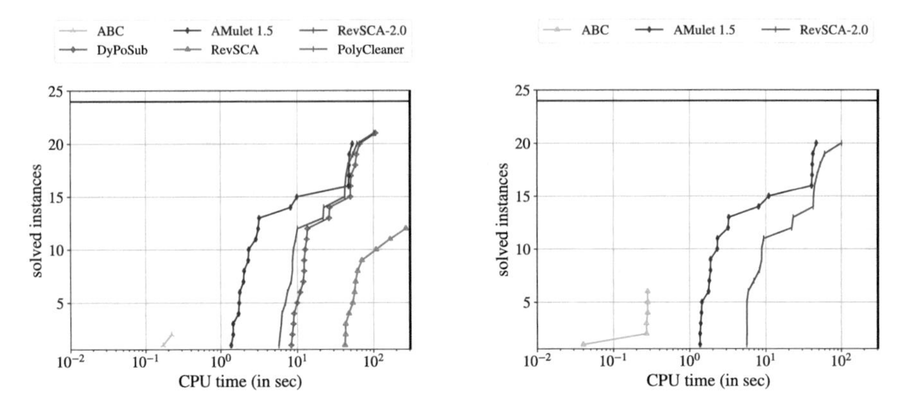

Fig. 4 Verification time (in sec) of unsigned (left) and signed (right) multipliers that are generated using GENMUL. Time limit: 300 s

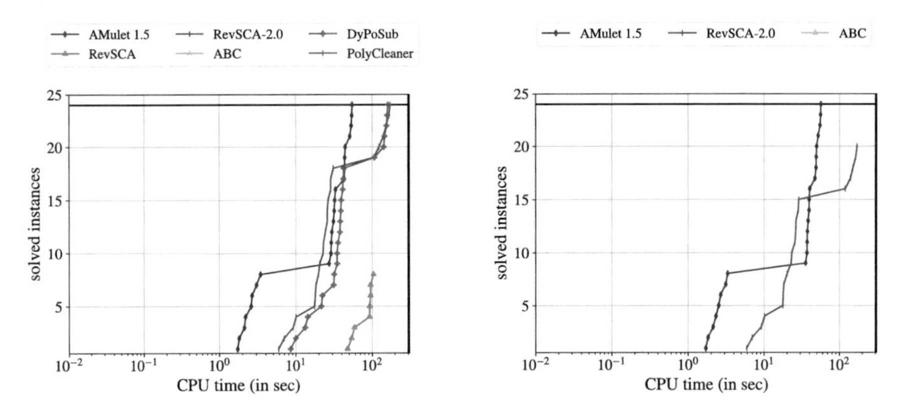

Fig. 5 Verification time (in sec) of unsigned (left) and signed (right) multipliers that are generated using MULTGEN. Time limit: 300 s

5.3 MULTGEN

In this experiment, we consider the 24 unsigned and 24 signed benchmarks of MULTGEN. We generated the benchmarks for an input bit-width of 64 and set the time limit to 300 s. The results are shown in Fig. 5.

The left side of Fig. 5 shows that AMULET 1.5, DYPOSUB, and REVSCA-2.0 are able to verify the complete benchmark set, with AMULET 1.5 being the fastest tool. Furthermore, AMULET 1.5 is able to verify all signed multipliers, cf. right side of Fig. 5. ABC and POLYCLEANER produce a segmentation fault or exceed the time limit for all instances.

5.4 EPFL Combinational Benchmark Suite

In this experiment, we verify the 64-bit multiplier that is contained in the EPFL Combinational Benchmark Suite. The time limit is set to 300 s and the results are shown in Table 1.

It can be seen that only DyPoSub and RevSCA-2.0 are able to verify this multiplier circuit. AMulet 1.5 and PolyCleaner exceed the time limit, ABC and RevSCA produce a segmentation fault.

5.5 ABC/Boolector

We generate simple array multipliers with an input bit-width of 128, 256, 512, and 1 024 using ABC and Boolector. Since both tools directly produce multipliers in the AIG format, we do not consider the tool PolyCleaner in this experiment. We set the time limit to 14 400 s (4 h), and the results can be seen in Fig. 6.

The multipliers generated by ABC/Boolector can be fully decomposed into half- and full-adders. Thus, we enable the optimization "&atree" in the verification tool ABC. As previously mentioned, the benchmarks generated by ABC and Boolector are internally equivalent, except the order of the inputs differs. In

Table 1 Verification time (in sec) of the multiplier contained in the EPFL Combinational Benchmark Suite

Tool	ABC	AMulet 1.5	DyPoSub	PolyCleaner	RevSCA	RevSCA-2.0
CPU time (in sec)	segfault	TO	59	TO	segfault	81

Time limit (TO): 300 s

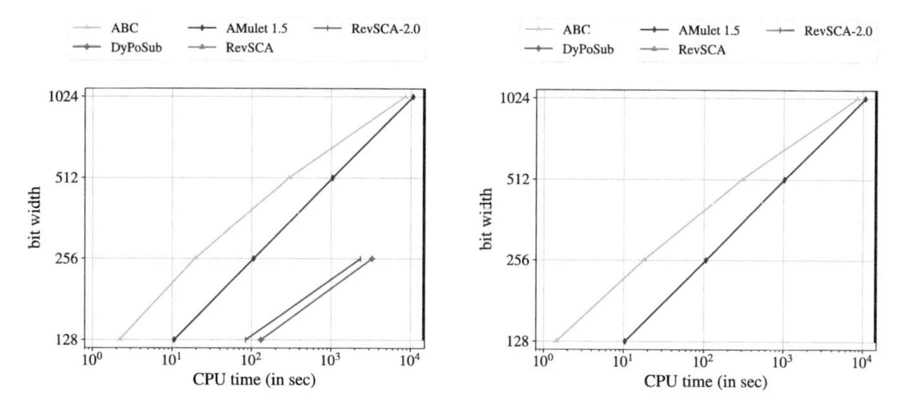

Fig. 6 Verification time (in sec) of simple array multipliers that are generated using ABC (left) and Boolector (right). Time limit: 14,400 s

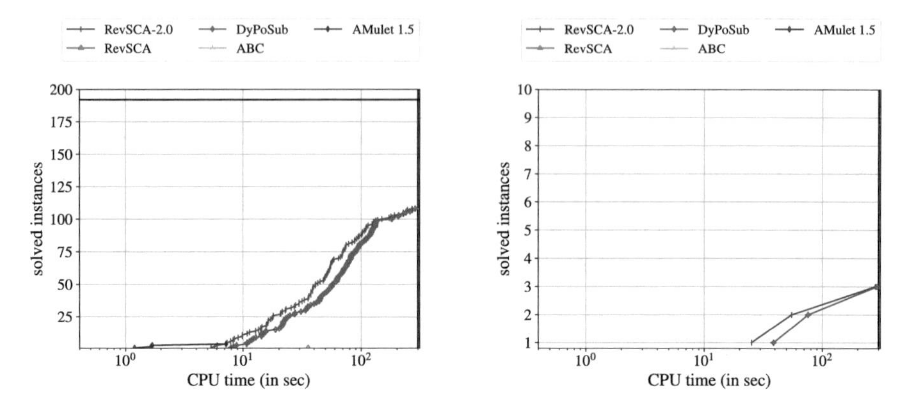

Fig. 7 Verification time of unsigned multipliers which are optimized using "resyn", without (left) and with technology mapping (right). Time limit: 300 s

ABC the order of inputs is $a_0, \ldots, a_{n-1}, b_0, \ldots, b_{n-1}$. In BOOLECTOR the order is $a_0, b_0, \ldots, a_{n-1}, b_{n-1}$.

It can be seen that ABC and AMULET 1.5 are able to verify all benchmarks, and ABC is slightly faster. The tools DYPOSUB and REVSCA-2.0 are only able to verify the ABC benchmarks, but produce wrong results, i.e., conclude that the multiplier is buggy, for multipliers generated by BOOLECTOR. REVSCA produces a segmentation fault for all experiments.

5.6 Optimized Benchmarks

In this experiment, we consider the 192 unsigned multiplier circuits that are generated by the ARITHMETIC MODULE GENERATOR. The time limit is set to 300 s. We optimize these benchmarks as described in Sect. 4.7 and either apply "resyn," "resyn3," or "dc2" in ABC. In all of these benchmark suites, we furthermore apply technology mapping using the standard cell library "mcnc.genlib" of SIS [48]. Thus, we gain six different setups, and the results can be seen in Figs. 7, 8, and 9.

It can be seen that DYPOSUB and REVSCA-2.0 are able to verify more than half of the benchmarks when only synthesis is applied. If additionally technology mapping is used to optimize the multipliers, these tools are only able to verify at most eight circuits. The tools AMULET 1.5 and REVSCA are able to verify a small amount (less than 10) of synthesized multipliers, but fail when technology mapping is applied. ABC produces a segmentation fault or an internal error for all experiments.

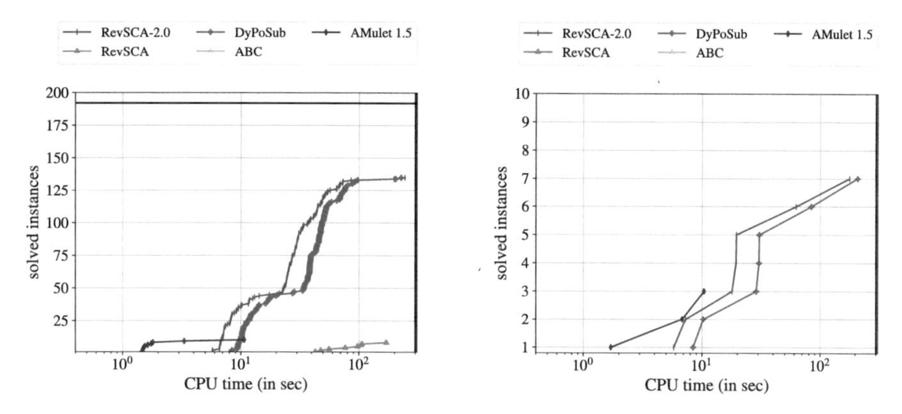

Fig. 8 Verification time of unsigned multipliers which are optimized using "resyn3", without (left) and with technology mapping (right). Time limit: 300 s

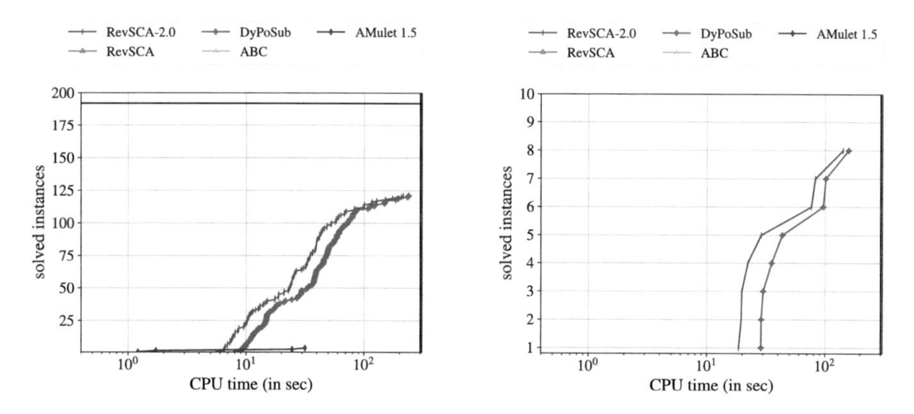

Fig. 9 Verification time of unsigned multipliers which are optimized using "dc2", without (left) and with technology mapping (right). Time limit: 300 s

5.7 *Proof Generation*

In our last experiment, we present the ability of generating proofs in AMULET 1.5. Again, we consider the 192 unsigned benchmarks of the ARITHMETIC MODULE GENERATOR and produce either PAC proofs or NSS proofs, which are then checked using PACHECK [30], PASTÈQUE [30], and NUSS-CHECKER [25]. The results are shown in Fig. 10, and it can be seen that generating proof certificates increases the computation time of AMULET 1.5. Both proof formats need the same amount of time for generation. In the right side of Fig. 10, we compare the checking time of the generated proofs. Proofs in the NSS format, which are checked by NUSS-CHECKER, are faster to check than PAC proofs that can be checked either by PACHECK or by PASTÈQUE. The verified proof checker PASTÈQUE is around four times slower than PACHECK.

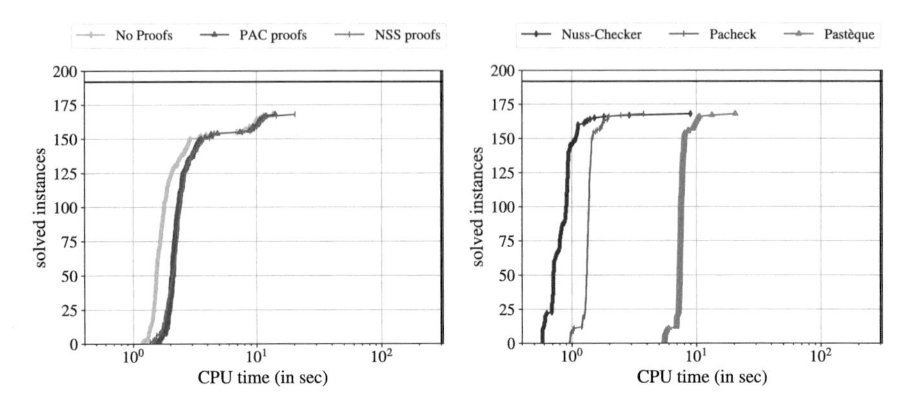

Fig. 10 Certification time of unsigned multipliers (left) and the proof checking time (right). Time limit: 300 s

6 Conclusion

In this chapter, we presented the current state of the art in verifying flattened gate-level integer multipliers using computer algebra. We introduced the verification technique and discussed the most recent work in this area. Tools that have been developed within the last 3 years were highlighted and publicly available benchmarks were presented. We concluded with a rigorous evaluation of the tools. Summarizing, no verification tool is a clear favorite over another. For simple multipliers, ABC [16, 60] is the fastest tool; however, ABC fails on complex multipliers. For non-optimized benchmarks, AMULET 1.5 [28], DYPOSUB [42], and REVSCA-2.0 [40] almost always solve around the same amount of benchmarks, where AMULET 1.5 is an order of magnitude faster than related work. Additionally AMULET 1.5 is the only tool which allows to generate proof certificates. DYPOSUB and REVSCA-2.0 outperform related work on optimized benchmarks, which are closer to industrial multipliers.

References

1. Amarú, L., Gaillardon, P.-E., De Micheli, G.: The EPFL combinational benchmark suite. In: International Workshop on Logic and Synthesis (IWLS), pp. 57–61 (2015)
2. Amarú, L., Gaillardon, P.-E., De Micheli, G.: The EPFL combinational benchmark suite (2020). https://www.epfl.ch/labs/lsi/page-102566-en-html/benchmarks/
3. Amarú, L., Gaillardon, P.-E., De Micheli, G.: The EPFL combinational benchmark suite (2020). https://github.com/lsils/benchmarks
4. Beame, P., Impagliazzo, R., Krajícek, J., Pitassi, T., Pudlák, P.: Lower Bounds on Hilbert's Nullstellensatz and Propositional Proofs. In: Proceedings of the London Mathematical Society, vol. s3–73, pp. 1–26 (1996)

5. Beame, P., Liew, V.: Towards Verifying Nonlinear Integer Arithmetic. In: Proceedings of the International Conference on Computer Aided Verification (CAV) 2017. LNCS, vol. 10427, pp. 238–258. Springer, New York (2017)
6. Berkeley Logic Synthesis and Verification Group. ABC: A System for Sequential Synthesis and Verification (2019). http://www.eecs.berkeley.edu/~alanmi/abc/. Bitbucket Version 1.01
7. Biere, A.: Collection of Combinational Arithmetic Miters Submitted to the SAT Competition 2016. In: SAT Competition 2016. Department of Computer Science Report Series B, vol. B-2016-1, pp. 65–66. University of Helsinki, Helsinki (2016)
8. Biere, A.: Weaknesses of CDCL solvers, August 2016. In: Fields Institute Workshop on Theoretical Foundations of SAT Solving. http://www.fields.utoronto.ca/talks/weaknesses-cdcl-solvers
9. Biere, A., Fazekas, K., Fleury, M., Heisinger, M.: CaDiCaL, Kissat, Paracooba, Plingeling and Treengeling entering the SAT Competition 2020. In: Proceedings of the SAT Competition 2020—Solver and Benchmark Descriptions. Department of Computer Science Report Series B, vol. B-2020-1, pp. 51–53. University of Helsinki, Helsinki (2020)
10. Bryant, R.E.: Graph-Based Algorithms for Boolean Function Manipulation. IEEE Trans. Comput. **35**(8), 677–691 (1986)
11. Bryant, R.E., Chen, Y.: Verification of arithmetic circuits using binary moment diagrams. STTT **3**(2), 137–155 (2001)
12. Buchberger, B.: Ein Algorithmus zum Auffinden der Basiselemente des Restklassenringes nach einem nulldimensionalen Polynomideal, PhD thesis. University of Innsbruck, Innsbruck (1965)
13. Chen, Y., Bryant, R.E.: Verification of Arithmetic Circuits with Binary Moment Diagrams. In: Design Automation Conference, DAC 1995, pp. 535–541. ACM, New York (1995)
14. Chen, Y., Clarke, E., Ho, P., Hoskote, Y., Kam, T., Khaira, M., O'Leary, J., Zhao, X.: Verification of All Circuits in a Floating-Point Unit Using Word-Level Model Checking. In: FMCAD 1996. LNCS, vol. 1166, pp. 19–33. Springer, Berlin (1996)
15. Ciesielski, M.J., Yu, C., Brown, W., Liu, D., Rossi, A.: Verification of Gate-level Arithmetic Circuits by Function Extraction. In: Design Automation Conference, DAC 2015, pp. 52:1–52:6. ACM, New York (2015)
16. Ciesielski, M.J., Su, T., Yasin, A., Yu, C.: Understanding algebraic rewriting for arithmetic circuit verification: a Bit-Flow model. IEEE TCAD, pp. 1–1 (2019). Early acces
17. Clegg, M., Edmonds, J., Impagliazzo, R.: Using the Groebner basis algorithm to find proofs of unsatisfiability. In: STOC 1996, pp. 174–183. ACM, New York (1996)
18. Cox, D., Little, J., O'Shea, D.: Ideals, Varieties, and Algorithms. Springer, New York (1997)
19. Heule, M.J.H., Biere, A.: Proofs for satisfiability problems. In: All about Proofs, Proofs for All Workshop, APPA 2014, vol. 55, pp. 1–22. College Publications, Australia (2015)
20. Homma, N., Watanabe, Y., Aoki, T., Higuchi, T.: Formal Design of Arithmetic Circuits Based on Arithmetic Description Language. IEICE Trans. **89-A**(12), 3500–3509 (2006)
21. Homma Laboratory, RIEC, Tohoku University. Arithmetic Module Generator. https://www.ecsis.riec.tohoku.ac.jp/topics/amg/
22. Hunt, W.A., Jr., Kaufmann, M., Moore, J.S., Slobodova, A.: Industrial hardware and software verification with ACL2. Philos. Trans. Royal Soc. A **375**(2104), 20150399 (2017)
23. Kaufmann, D.: AMulet 1.5 (2020). https://github.com/d-kfmnn/amulet
24. Kaufmann, D.: Formal Verification of Multiplier Circuits using Computer Algebra. PhD thesis, Informatik, Johannes Kepler University Linz (2020)
25. Kaufmann, D., Biere, A.: Nullstellensatz-proofs for multiplier verification. In: CASC. Lecture Notes in Computer Science. Springer, Berlin (2020, to appear)
26. Kaufmann, M., Moore, J.S.: ACL2 Version 8.2 (2019). http://www.cs.utexas.edu/users/moore/acl2/
27. Kaufmann, D., Biere, A., Kauers, M.: Incremental Column-wise verification of arithmetic circuits using computer algebra. In: Formal Methods in System Design (2019). Online first
28. Kaufmann, D., Biere, A., Kauers, M.: Verifying Large Multipliers by Combining SAT and Computer Algebra. In: FMCAD 2019, pp. 28–36. IEEE, New York (2019)

29. Kaufmann, D., Kauers, M., Biere, A., Cok, D.: Arithmetic Verification Problems Submitted to the SAT Race 2019. In: SAT Race 2019. Department of Computer Science Report Series B, vol. B-2019-1, p. 49. University of Helsinki, Helsinki (2019)
30. Kaufmann, D., Fleury, M., Biere, A.: Pacheck and Pastèque, Checking Practical Algebraic Calculus Proofs. In: FMCAD 2020. FMCAD, vol. 1, pp. 264–269. TU Vienna Academic Press, Austria (2020)
31. Kuehlmann, A., Paruthi, V., Krohm, F., Ganai, M.: Robust Boolean reasoning for equivalence checking and functional property verification. IEEE TCAD **21**(12), 1377–1394 (2002)
32. Liew, V., Beame, P., Devriendt, J., Elffers, J., Nordström, J.: Verifying Properties of Bit-vector Multiplication Using Cutting Planes Reasoning. In: FMCAD 2020. FMCAD, vol. 1, pp. 194–204. TU Vienna Academic Press, Austria (2020)
33. Lv, J., Kalla, P.: Formal Verification of Galois Field Multipliers Using Computer Algebra Techniques. In: International Conference on VLSI Design, VLSID 2012, pp. 388–393. IEEE Computer Society, New York (2012)
34. Lv, J., Kalla, P., Enescu, F.: Efficient Gröbner basis reductions for formal verification of Galois field arithmetic circuits. IEEE TCAD **32**(9), 1409–1420 (2013)
35. Mahzoon, A., Große, D., Drechsler, R.: GenMul (2018). http://sca-verification.org/genmul
36. Mahzoon, A., Große, D., Drechsler R.: GenMul (2018). https://github.com/amahzoon/genmul
37. Mahzoon, A., Große, D., Drechsler, R.: PolyCleaner (2018). http://sca-verification.org/polycleaner
38. Mahzoon, A., Große, D., Drechsler, R.: PolyCleaner: clean your polynomials before backward rewriting to verify Million-gate Multipliers. In: ICCAD 2018, pp. 129:1–129:8. ACM, New York (2018)
39. Mahzoon, A., Große, D., Drechsler, R.: RevSCA and RevSCA-2.0 (2019). http://sca-verification.org/revsca
40. Mahzoon, A., Große, D., Drechsler, R.: RevSCA: using reverse engineering to bring light into backward rewriting for big and dirty multipliers. In: DAC 2019, pp. 185:1–185:6. ACM, New York (2019)
41. Mahzoon, A., Große, D., Scholl, C., Drechsler, R.: DyPoSub (2020). http://sca-verification.org/dyposub
42. Mahzoon, A., Große, C. Scholl, D., Drechsler, R.: Towards formal verification of optimized and industrial multipliers. In: DATE, pp. 544–549. IEEE, New York (2020)
43. Niemetz, A., Preiner, M., Wolf, C., Biere, A.: Btor2, BtorMC and Boolector 3.0. In: CAV 2018. LNCS, vol. 10981, pp. 587–595. Springer, Berlin (2018)
44. Parhami, B.: Computer Arithmetic—Algorithms and Hardware designs. Oxford University, Oxford (2000)
45. Pavlenko, E., Wedler, M., Stoffel, D., Kunz, W., Wienand, O., Karibaev, E.: Modeling of custom-designed arithmetic components for ABL normalization. In: Forum on Specification and Design Languages, FDL 2008, pp. 124–129. IEEE, New York (2008)
46. Ritirc, D., Biere, A., Kauers, M.: A practical polynomial calculus for arithmetic circuit verification. In: SC2 2018, pp. 61–76. CEUR-WS (2018)
47. Sentovich, E., Singh, K., Lavagno, L., Moon, C., Murgai, R., Saldanha, A., Savoj, H., Stephan, P., Brayton, R.K., Sangiovanni-Vincentelli, A.L.: SIS. https://ptolemy.berkeley.edu/projects/embedded/pubs/downloads/sis/index.htm
48. Sentovich, E., Singh, K., Lavagno, L., Moon, C., Murgai, R., Saldanha, A., Savoj, H., Stephan, P., Brayton, R.K., Sangiovanni-Vincentelli, A.L.: SIS: a system for sequential circuit synthesis. Technical Report UCB/ERL M92/41, EECS Department, University of California, Berkeley (1992)
49. Sharangpani, H., Barton, M.L.: Statistical Analysis of Floating Point Flaw in the Pentium Processor (1994)
50. Stoffel, D., Kunz, W.: Equivalence checking of arithmetic circuits on the arithmetic bit level. IEEE TCAD **23**(5), 586–597 (2004)
51. Temel, M.: MultGen (2020). https://github.com/temelmertcan/multgen

52. Temel, M., Slobodová, A., Hunt, W.A.: Automated and scalable verification of integer multipliers. In: CAV (1). Lecture Notes in Computer Science, vol. 12224, pp. 485–507. Springer, Berlin (2020)
53. Vasudevan, S., Viswanath, V., Sumners, R.W., Abraham, J.A.: Automatic verification of arithmetic circuits in RTL using stepwise refinement of term rewriting systems. IEEE Trans. Comput. **56**(10), 1401–1414 (2007)
54. Wienand, O., Wedler, M., Stoffel, D., Kunz, W., Greuel, G.: An algebraic approach for proving data correctness in arithmetic data paths. In: International Conference on Computer Aided Verification, CAV 2008. LNCS, vol. 5123, pp. 473–486. Springer, Berlin (2008)
55. Wolf, C.: Yosys Open SYnthesis Suite. http://www.clifford.at/yosys/
56. Yu, C.: Algebraic RewriTing in ABC (2018). https://github.com/ycunxi/abc
57. Yu, C., Brown, W., Liu, D., Rossi, A., Ciesielski, M.J.: Formal verification of arithmetic circuits by function extraction. IEEE TCAD **35**(12), 2131–2142 (2016)
58. Yu, C., Ciesielski, M.J.: Efficient parallel verification of galois field multipliers. In: Asia and South Pacific Design Automation Conference, ASP-DAC 2017, pp. 238–243. IEEE, New York (2017)
59. Yu, C., Ciesielski, M.J.: Formal analysis of Galois field arithmetic circuits-parallel verification and reverse engineering. IEEE TCAD **38**(2), 354–365 (2019)
60. Yu, C., Ciesielski, M.J., Mishchenko, A.: Fast algebraic rewriting based on and-inverter graphs. IEEE TCAD **37**(9), 1907–1911 (2018)

The Vital Role of Machine Learning in Developing Emerging Technologies

Victor M. van Santen, Florian Klemme, and Hussam Amrouch

1 Introduction

Circuit simulations are the key to the evaluation of a semiconductor technology at the circuit level. The performance, power, and efficiency of a specific circuit design in a specific technology are evaluated in circuit simulators such as SPICE. The transistors in such circuit simulators are implemented through compact models such as BSIM-CMG [9, 35]. Transistor compact models are an abstracted high-performance implementation of the electrical behavior of a transistor. A semiconductor manufacturer (foundry) can then use a set of parameters called a *model card* to calibrate this behavior to the observed behavior (e.g., based on prototype measurements) of a specific technology (e.g., 22 nm FinFET). However, this standard approach of circuit simulations faces three key challenges.

Foundry Secrecy First, the foundries are reluctant to share transistor model cards, as this could allow reverse engineering of their commercial product. For example, the frequently used *arbitrary units* (a.u.) across publications [11, 36, 38] and the Non-Disclosure-Agreements (NDA) guarding Process Design Kits (PDK) access protect their intellectual property.

Innovation Requires New Models Secondly, with the end of Dennard scaling and the current age of innovation in semiconductor technologies, transistors are not just geometrically scaled anymore but instead altered in their fundamental structure. Innovations in recent times were the introduction of high-k metal gates in 32 nm technology and 22 nm introducing 3D FinFET. For compact models, this innovation poses a serious challenge. With each innovation, it requires a new

V. M. van Santen (✉) · F. Klemme · H. Amrouch
University of Stuttgart, Stuttgart, Germany
e-mail: van-santen@iti.uni-stuttgart.de; klemme@iti.uni-stuttgart.de; amrouch@iti.uni-stuttgart.de

R. Drechsler, D. Große (eds.), *Recent Findings in Boolean Techniques*,
https://doi.org/10.1007/978-3-030-68071-8_2

compact model instead of updated parameters in model cards as the underlying principles of the transistor change (e.g., introducing BSIM-CMG [9] for FinFET over BSIMv4 [24] for MOSFET). However, compact model development can take years. Development can only start once the represented technology is mature and its physics are understood. This delay in modeling availability hinders the use of standard EDA tools to evaluate circuit designs in such new technologies, which hinders market entry and prolongs time to market for the emerging technology.

Early Evaluation of Technology Thirdly, various competing transistor technologies exist simultaneously. For example, FinFET currently competes against FDSOI, Nanowire, and Nanosheet transistors in traditional CMOS and NC-FinFET, TFET, and other transistors in emerging technologies. Currently, each technology has prototypes that are measured as well as material and physics simulations in TCAD. However, in order to evaluate if a particular transistor technology is suitable for a specific circuit currently in the design phase, we need support within the EDA standard flow (e.g., in circuit simulators). Therefore, traditional evaluation mandates the development of various compact models and the calibration in various model cards. This is a complex and thus costly and labor-intensive process, as transistors are calibrated for different voltages, temperature, variants (high-Vth, low-Vth, high power, etc.), and geometries to just name a few variables. Therefore, before a technology can be evaluated to determine commercial or academic interest, considerable investments are necessary to develop and calibrate compact models.

1.1 Machine Learning Transistor Model

For the transistor modeling in this work, we propose Machine Learning (ML), specifically Neural Network (NN)-based transistor models to act as an intermediate between early data from prototypes and complex compact models. Our goal is to provide NN-based transistor models, tackling the three challenges outlined above. First, NN-based transistor models are by its very nature black-box modeling approaches and thus cannot expose manufacturing details about the transistors, which protects the intellectual property of the foundries. Furthermore, NN-based transistor models are generic, i.e., a single approach can apply to different transistor technologies (as we show in Sect. 4.8) and thus do not require continuous development of new models. Lastly, NN-based modeling can predict without grasping the underlying fundamentals (physics, materials, etc.), enabling quick development of these NN-based models despite transistor innovation. Naturally, an intermediate NN-based model is trained on limited data sets and thus less accurate than a fully calibrated and developed compact model (i.e., the model does not replace the end product).

1.2 Machine Learning Standard Cell Model

For analogue circuits, SPICE-based circuit simulations, which rely on transistor models, are used. However, for large-scale digital circuits, standard cells are the industry standard approach for circuit design. Standard cell libraries are the established way to share process technology between the foundry and circuit designers. They provide a description of cells, suitable to be used by other tools in the Electronic Design Automation (EDA) flow. Hence, for the same three challenges (secrecy, transistor innovation, and early designs), a transistor-level model is insufficient.

In this work, we propose a solution that enables the designer to design the circuit without the need of hard-to-acquire files or time-consuming standard cell library characterizations. To achieve this, we replace the traditional cell library depicted in Fig. 1 with a Machine Learning (ML) approach that quickly generates cell libraries on demand (shown in Fig. 2). Similarly, to transistor modeling per ML, the ML-based models are not perfectly accurate. Instead, sufficient accuracy to reveal tendencies and enabling the designer to perform design space exploration is the goal. Due to the vast increase in generation speed compared to traditional characterization, we can use established optimization algorithms to search the

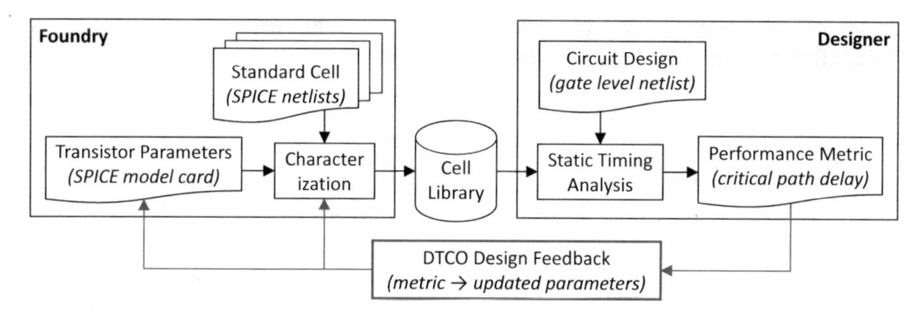

Fig. 1 Traditional DTCO flow. Circuit-dependent performance results are fed back to fine-tune transistor and characterization parameters

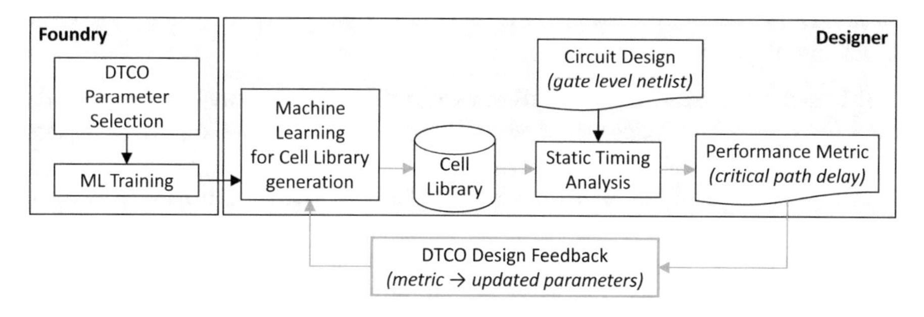

Fig. 2 Our machine learning approach removes the foundry from the DTCO feedback loop

design space. Then, when foundry access is indeed achieved, the results can be verified with accurate, yet time-consuming SPICE-based library characterization.

Additionally, Design Technology Co-Optimization (DTCO) can be used to optimize technology and circuit design side by side. As depicted in Fig. 1, DTCO extends the traditional design flow by adding a feedback loop to the process. The cell library and the target circuit design are evaluated together, and a corresponding performance metric, e.g., the critical path delay, is calculated and fed back. This way, transistor parameters are optimized in each iteration to achieve the best performance on the circuit.

In practice, DTCO is frequently faced with the three challenges outlined before. To make the iterative process feasible, all tools and files that are part of the cycle have to be accessible by one party. However, both transistor and cell data are confidential. In addition, for emerging technologies, physics-based compact models may not be available yet. On top of the challenges for transistor models, the process of cell library characterization is time-consuming, requiring thousands of SPICE simulations. Due to all these challenges, DTCO can solely be performed by a select group of circuit designers (e.g., designers with non-disclosure agreements with foundries).

Instead, our approach takes available transistor and characterization parameters as input and quickly predicts a corresponding library. Should the transistor parameters be confidential, then our NN-based model could provide them without disclosing information. This eliminates the foundry and provides a shortcut in the design feedback loop, i.e., decoupling foundry and designer.

2 Related Work

ML-Based Transistor Models with Domain Knowledge Early works in ML-based transistor modeling had limited neurons and layers due to limitations in computational power and tool support of the time [19, 37]. Therefore, various works used domain knowledge to augment the NN and improve its accuracy. However, it cannot be applied to emerging technologies (which are not yet fully understood) and customize (as well as constraints) the model for a particular technology (removes generality of the model).

ML-Based Transistor Models Without Prior Knowledge Approaches which do not rely on prior knowledge can be generic (apply to many technologies) but require more computational resources.

A combination of NN and genetic algorithms is used in [16]. In addition to the regular training, the genetic algorithm is used to find the best structure of the NN. However, only regular planar MOSFET transistors are evaluated and only the current is inferred. Hence, the work lacks information if this could be applied to emerging technologies is missing as well as a sufficient validation (see Sect. 4.6 for details).

Zhang et al. [38] present an NN-based surrogate model as a compact model alternative for novel transistors. They demonstrate their framework for FinFET as an established technology and TFET as an emerging technology with no available compact model. However, prediction accuracy is not discussed on transistor I–V curves. Therefore, their validation is lacking (see Sect. 4.6) and we cannot estimate their achieved modeling accuracy with respect to the transistor parameters (Vth, Ion, Ioff, etc.).

ML-Based Standard Cell Models In [30], linear regression is used to predict the delay of a single standard cell. As input features, they consider current and voltage on cell terminals at switching time as well as some transistor parameters. The paper is probably the closest to our work in regards to cell prediction; however, their choice of input features makes their approach unsuitable for our problem. Having voltages as input features would require further simulations to generate results on circuit designs. Likewise, the impact on circuit performance is not considered in their work, instead, they lay focus on a single DFF cell.

An ML-for-DTCO approach is found in [38]. Instead of predicting standard cells, they use ML to replace the compact model. Although this also helps with long simulation times, they focus on reducing the overhead of developing a compact model.

Ceyhan et al. [8] apply ML for design, technology, and ingredient (DTI) optimization. Although standard cell libraries are considered in their work, they do not generate them by ML. Instead, they consider different libraries alongside other inputs to optimize the design in a holistic, zoomed-out view.

3 Background

3.1 Compact Models

Compact models bring a transistor implementation into the SPICE simulator. They abstract the underlying physical equations to electrical behavior. They provide high accuracy with the minimum number of equations and parameters (to improve performance). Actual physical modeling of the transistors materials is done in TCAD simulators, which are very computational complex and thus time-consuming. Unfortunately, these TCAD simulation cannot be used for circuit simulations due to their computational complexity [10]. Typically TCAD simulations (simulated transfer curves) and selected experimental data sets (measured transfer curves) are used to calibrate the parameters (the transistor model card) of a particular transistor (e.g., Intel FinFET) to a particular transistor model (e.g., BSIM-CMG). BSIM-CMG has over 100 parameters in the model card, spanning geometry, dopant concentrations, material, and electrical properties [9]. Therefore, a compact model

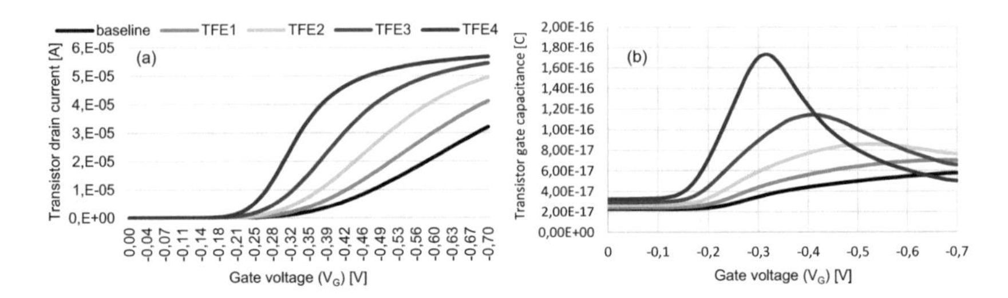

Fig. 3 (**a**) Shows the I-V characteristics of Negative Capacitance Transistors. The thicker the ferroelectric thickness, the higher the on-current. However, this comes at the cost of a noticeable increase in the gate capacitance C_{GG}

can at its earliest be developed as soon as the physics are fully understood and the technology reached sufficient maturity and a TCAD model is available.

3.2 *Negative Capacitance FinFET*

NC-FinFET is an emerging transistor technology with a ferroelectric layer integrated into the gate stack of a FinFET transistor [28]. The ferroelectric layer provides a voltage amplification of the gate voltage applied to the transistor, and thus changes the electrical behavior of the FinFET. For example, its sub-threshold slope is below 60 mV/decade, making NC-FinFET a promising technology to succeed FinFET. The sub-threshold slope is an important characteristic to determine the switching speed of a transistor. A steeper sloped can provide a higher on-currents or less leakage when the transistor is turned off. Sub-threshold slopes of both FinFET and NC-FinFET are shown in Fig. 3a along with the resulting increase in the total gate capacitance of transistor in (b).

Another characteristic of NC-FinFET is reduced short channel effects due to a low or even negative Drain-Induced Barrier Lowering (DIBL) effect [3]. Additionally, the ferroelectric gate increases overall gate capacitance C_{gg}. With the voltage-dependent voltage amplification, negative DIBL, C_{gg}-increase, and steep sub-threshold slope, NC-FinFET shows an increased complexity in its behavior compared to conventional FinFET. Therefore, it is a prime candidate to test if our approach is applicable to emerging technologies without modifications.

3.3 *Transistor Characteristics*

Transistor characteristics are used to evaluate and compare the performance of the transistor with regard to certain properties. Due to their importance in evaluation,

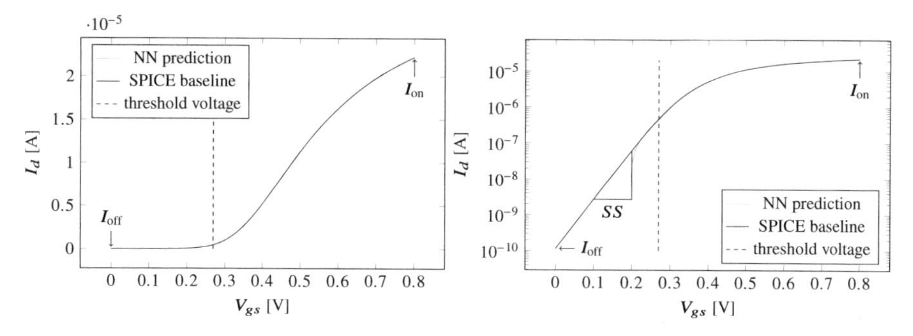

Fig. 4 I_d-V_{gs} curve with $V_{ds} = 0.1\,\text{V}$. To determine the sub-threshold slope (SS) visually, the plot is repeated in logarithmic scale on the right side

we use the following transistor characteristics as additional accuracy metrics in our ML approach.

I_{off} the off-current at $V_{gs} = 0\,\text{V}$,
I_{on} the on-current at $V_{gs} = V_{DD}$,
V_{th} the threshold voltage, i.e., voltage at which the transistor turns on or off,
SS the sub-threshold slope, i.e., fast a transistor turns on or off.

These transistor characteristics are also visualized in Figs. 3 and 4.

4 Our Machine Learning Transistor Model

In this work, we propose an NN to learn and reproduce the I–V curve (electrical response) of a transistor. If the NN-based transistor model can accurately predict the drain current of a transistor, then it is sufficient to be employed in SPICE simulations. SPICE model circuit components (resistors, capacitors, transistors, diodes, etc.) based on their conduction (inverse resistance) from each terminal (gate G, source S, drain D, bulk B for transistors) to each other terminal (for details on SPICE solving see our previous work [34]). Since SPICE knows the voltage at the terminals and the NN-based model provides the currents, the conductances can be calculated with $G = \frac{1}{R} = \frac{I}{V}$ [34]. Therefore, for the rest of this work, we solely discuss the I–V curve, also called the transfer curve.

4.1 Experimental Setup

Our NN-based transistor model is a prototype application in C++ using PyTorch for ML functionality. We use a fully connected feed-forward NN with 2 layers, 500 nodes in each layer, and PReLU as the activation function. The NN is trained using back-propagation with stochastic gradient descent. As established in ML, input values are normalized before training and inference, so that an adaptation of the learning rate to the training data is not required. For validation, 30% of the training data is preserved to be used as the validation data set.

For the generation of training data as well as for validation, existing compact models are used. BSIM-CMG [9, 35] is used for conventional FinFET. A compact model [22] based on BSIM-CMG is used for emerging NC-FinFET. Details on the used NC-FinFET modeling and FinFET device calibration with industrial measurements are available in [5] and [20], respectively.

The hyperparameters of the NN are optimized based on the training data from the conventional FinFET model, as commonly done in ML. However, the NN has been purposely up-scaled slightly (to 2 layers and 500 nodes each), to increase the chance of better adaptability to different transistors. To allow a comparison of our work against previous works, we express accuracy in the traditional R^2 score, as well as our own metrics. The R^2 score expresses the mean error of the prediction in relation to the total variance. The R^2 score is defined as

$$R^2 = 1 - \frac{\sum (Y_{\text{true}} - Y_{\text{pred}})^2}{\sum (Y_{\text{true}} - \bar{Y}_{\text{true}})^2} \tag{1}$$

where Y_{pred} is the predicted value, Y_{true} is the actual value (from the test set), and $\bar{Y}_{\text{true}}$ is the mean value of the test set. An R^2 score of 1 indicates perfect accuracy, whereas a score around 0 represents randomly guessing a value around the mean. The interpolation of two individual NNs as described in the next section is implemented with a custom Python script, first invoking the individual NNs and interpolating the results afterward.

4.2 *Data Scaling*

Modeling the transfer curve is a challenge since this I–V curve spans multiple orders of magnitude in terms of currents (small leakage current, yet million times stronger drive currents). Therefore, applying standard ML techniques when using the mean squared error (MSE) as the fitness function during training is problematic. While this works fine for large current values, smaller values exhibit high relative errors. The MSE value is dominated by the mismatch in the high-value region and the errors in small values are not weighted enough. However, key transistor characteristics like the sub-threshold slope and leakage current I_{off} is determined within this lower-value range. Thus, key transistor parameters are susceptible to error due to inadequate training. This is a common problem also faced by other works [19]. Figure 4 illustrates this problem. In the linear representation on the left side, the I–V curves of SPICE and the NN clearly overlap. In the logarithmic plot on the right side, the problem becomes apparent: For lower values, the NN curve

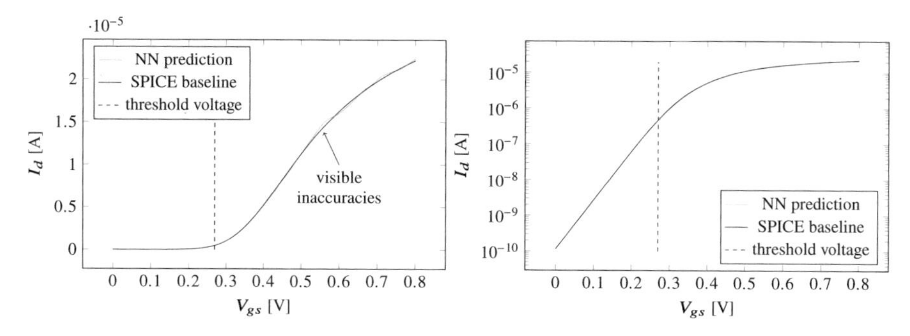

Fig. 5 With logarithmically scaled training data, the I–V curves show a good match. However, at linear scale, inaccuracies appear at higher values

diverges from the SPICE baseline, eventually disappearing when negative values are predicted. The sub-threshold slope and leakage current cannot be derived properly.

Our solution to this problem is to scale the data before training. By applying a logarithmic scaling to the training data, the range of values is more equally weighted, i.e., formerly small values are taken more into account. Unfortunately, errors on formerly large values are less considered. Thus, the linear representation of the data shows similar high relative errors for the large values, as shown in Fig. 5.

To solve this issue and get acceptable error figures both small and large values, a first approach is to use logarithmic scaling only for the values below the threshold voltage and linear (i.e., no) scaling for the rest. To connect the two different ranges, the normalization is split into two parts, too. The logarithmic part is normalized in the interval $[-1; 0]$ and the linear part is normalized in the interval $[0; 1]$. This split of the normalization range prevents overlapping input ranges (due to different scaling) and prevents non-determinism in the training data (see Fig. 6). However, with the two scalings, the NN has problems approximating the function around the threshold voltage (V_{th}) where the differently scaled value ranges connect. From the perspective of the NN, the function features a discontinuity at V_{th} (connecting the two data sets), which leads to different values compared to the SPICE baseline right at V_{th}.

To make the mixed-scaling work and resolve inaccuracies at the corners, the NN could be duplicated with an additional node that decides which NN to query, depending on the input. However, an easier way is to use two different NNs independently and interpolate the results outside of the NNs. This approach can be observed in Fig. 7. The interpolation takes place in the green highlighted area (not V_{th}) which is at $V_{gs} = 0.4\,\text{V}$ with a width of $0.1\,\text{V}$. With this configuration, the difference between the combined output of the NNs and the baseline output from SPICE is minimized.

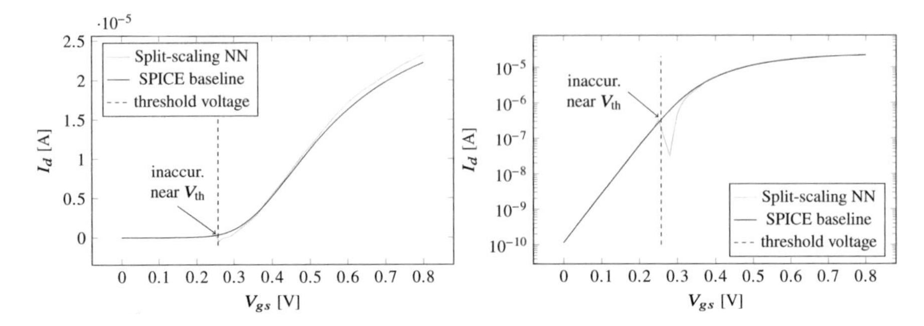

Fig. 6 The I–V curve with split-scaling around the threshold voltage. Hard to see: In the lower-value range, both curves match quite well. However, there are obvious errors around the threshold voltage

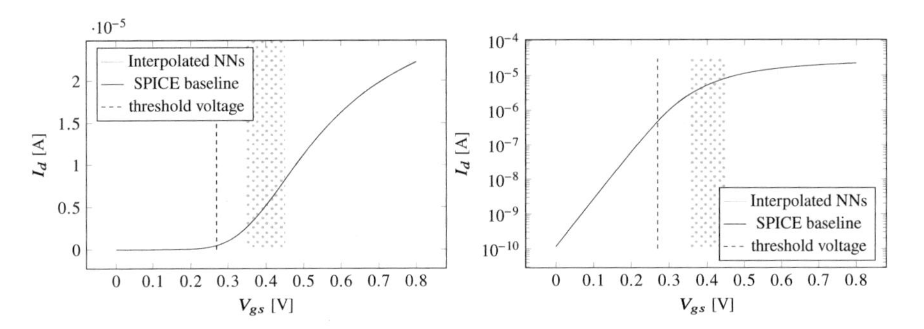

Fig. 7 Interpolation of two independently trained NNs

4.3 Advantages of NN-Based Transistor Modeling

Using ML techniques to model the behavior of a transistor has multiple advantages. First and foremost, only exemplary measurement data is needed to train the NN. No details about manufacturing or underlying physics are required. This is a core principle of ML, however, especially helpful for emerging technologies as their underlying physics are often still investigated and not yet fully understood.

Domain Knowledge Many ML applications incorporate domain knowledge in their structures and algorithms. Forcing an NN to follow certain assumptions about the transistor can help improving stability, training time, and correctness. However, falsely applied assumptions can make it impossible for the NN to fit certain data. For example, looking at conventional FinFET, one would assume monotonicity in the I_d-V_d curve. However, novel NC-FinFET experiences a negative DIBL effect that breaks this assumption (shown in Fig. 9). A strictly monotonic NN would not be able to learn this behavior properly. Therefore, to provide a generic NN-based transistor model, applicable to multiple technologies, no domain knowledge can be used to augment the NN.

Development Time While the development time of sophisticated compact models is usually in the order of years, ML training can be done within hours. As soon as early measurements from silicon become available, the NN can be trained. However, a certain amount of data is required to achieve meaningful accuracy. TCAD simulations can support measured data, especially when larger data sets under varying conditions (temperature, voltages, etc.) should be trained for. In this work, for simplicity but without loss of generality, we take an existing compact model as the source for training data and validation.

Secrecy of Neural Networks An NN is generally conceived as a black-box function. Even with the weights at hand, the contribution of individual parts to the output is often unclear. In fact, comprehending and tracing the decisions of an NN is a challenge and a research topic in its own [7]. Moreover, the structure of the NN does not reflect the structure of the mathematical equations of the underlying physics. Therefore, even if contributions would be understood, all variables are arbitrary parameters with no traceable relation to the real technology parameters such as geometries or dopant concentrations. Also, unlike in traditional compact modeling, there is no separation of technology (compact model) and calibration (model card) in the NN. The transistor calibration is implicit and indistinguishable from the fundamental technology. This adds an additional layer of abstraction to the NN model. *With all this complexity and abstractions in the NN, the extraction of technology details is at least unfeasible, if not impossible.* This offers great advantages in confidentiality and secrecy with regard to emerging and commercial technologies. NN-based transistor models can be shared without the risk of leaking confidential information or reverse engineering.

4.4 Disadvantages of NN-Based Transistor Modeling

Unfortunately, NN-based transistor models also feature some considerable disadvantages, i.e., cannot fully replace compact models and only act as an intermediate.

Scalability A noteworthy disadvantage of ML is the prediction accuracy outside of the range of the training data. Compact models are built on physical equations which allows them to be scalable beyond available data from silicon [10]. For example, with calibration between 20 and 50 °C, the compact model would provide decent accuracy at 80 °C, while the NN-based approach would struggle severely. Therefore, NN-based approaches should be trained for wide ranges to ensure that the inference always occurs within the trained range. Additionally, this highlights why NN-based transistor modeling is just an intermediate before the availability of actual compact models for the technology.

Learning Challenges Although the introduction of linear/log mixed-scaling in Sect. 4.2 improved the accuracy of the NN, the logarithmic scaling also created a new issue. Comparing the training effort for data on linear and logarithmic scaling,

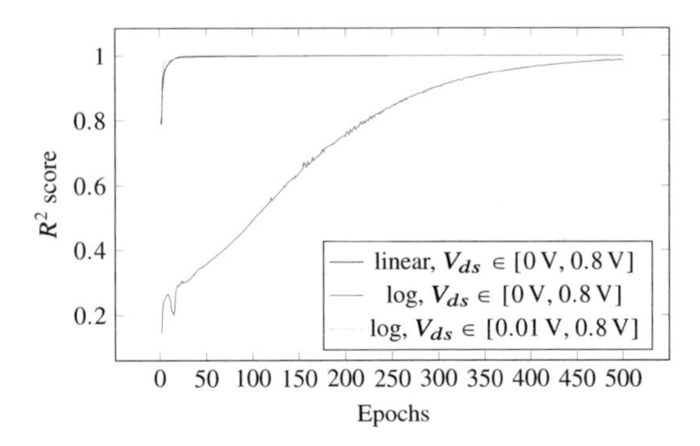

Fig. 8 Learning curves for different scalings. As soon as $V_{ds} = 0\,\text{V}$ is removed from the logarithmic data set, the learning curve improves significantly

a big discrepancy can be observed in Fig. 8. This is caused by troublesome data in the training set. For $V_{ds} = 0\,\text{V}$, SPICE reports a current of exactly 0 A. Since $\log 0$ is not defined, all values $\leq 10^{-30}$ are set to 10^{-30} during scaling. From a logarithmic point of view, this creates a gap between values with $V_{ds} = 0\,\text{V}$ and $V_{ds} = 0.01\,\text{V}$ which is perceived as a discontinuity by the NN. As a workaround, all data with $V_{ds} = 0\,\text{V}$ could be removed from training and test sets. As we can see in Fig. 8, the learning curve improves greatly as soon as these data points are removed.

4.5 *Inference Accuracy and Training Time*

Training data is generated by sweeping input parameters with a certain step size in SPICE simulations. All input parameter permutations and their corresponding outputs are collected in CSV files to serve as training and validation data sets for the NN. After 500 epochs of training, the R^2 scores for the training set, test set, and complete data set validation are 0.999772, 0.999764, and 0.999769, respectively. The R^2 score of the training set is very close to 1, meaning that the NN was able to match the training data excellently. Likewise, the R^2 score of the test set is only slightly lower than the R^2 score of the training set which indicates good generalization of the NN to the learned function (no overfitting).

Table 1 shows the R^2 score of different NNs after 48 h of training. Depending on the number of input and output parameters, the required training time changes. Without the number of fins in the FinFET (n_{fin}) configurable, NN 1 reaches an R^2 score of 0.999 for all output parameters with both linear and logarithmic scaling.

But when adding number of fins (n_{fin}) as an additional input parameter, the precision significantly drops, most notably for I_d with the linear scaling and an R^2 score of 0.786 (highlighted in Table 1). The reason behind this is the large amount of input data, which, in this case, contains 24 different temperatures (20 °C–135 °C), 80 different V_{ds} steps (0.01 V–0.8 V), 81 different V_{gs} steps (0 V–0.8 V), and 7 different number of fins (1–7). This accumulates to a total number of 1,088,640 data points, resulting in a much longer training time for a single epoch.

Table 1 R^2 scores for different NNs

Scale		Num. of steps T	V_{ds}	n_{fin}	R^2 score of output parameters I_d	c_{gb}	c_{gs}	c_{gd}
NN 1	lin	24	80	–	0.9999	0.9997	0.9997	0.9993
	log	24	80	–	0.9998	0.9993	0.9998	0.9989
NN 2	lin	24	80	7	0.7863	0.9894	0.9626	0.9834
	log	24	80	7	0.9856	0.9448	0.9265	0.9072
NN 3	lin	6	12	7	0.9996	0.9997	0.9997	0.9998
	log	6	12	7	0.9989	0.9995	0.9996	0.9996

Including n_{fin} increases training effort and thus, R^2 score for a given training time

While longer training times can be overcome with more processing power or just waiting longer, longer training cannot guarantee continuous improvement (i.e., converges to a low R^2 score). Alternatively, the training data sets can be trimmed. In our case, there is plenty of training data and a reduction shows no negative impact. NN 3 in Table 1 shows the results of using only 6 temperature steps and 12 V_{ds} steps for the training. With this optimization, the NN is capable to learn all 4 output parameters depending on all 4 input parameters within 48 h of training time.

4.6 *Traditional Fitness Compared to Transistor Metric Fitness*

The R^2 score metric provides an impression of the accuracy of an NN. However, inaccurate regions (outliers) might be overshadowed by good overall accuracy. Typically, this is not an issue, as overall accuracy is key. However, for transistors, certain transistor characteristics are critical for the evaluation of the transistor in circuit simulation. For example, the leakage current at $V_{gs} = 0V$ and the on-current at $V_{gs} = V_{DD}$ are ends of the I–V curve, but very important since digital circuits operate a majority of the time in these two extremes. Therefore, inaccurate modeling at these locations might severely alter the circuit simulation results of digital circuits and as such an R^2 score infers little about circuit simulation accuracy of a transistor model. For this reason, we propose to determine key transistor characteristics instead and use this as our metric to decide after how many training epochs a sufficient accuracy is reached. The comparison between R^2 score and transistor metric fitness is shown in Table 3.

For the training set, temperature ranges from 20 °C to 135 °C, V_{ds} from 0.01 V to 0.8 V, and V_{gs} from 0 V to 0.8 V. For the following tests, we use an enlarged test set with generated data at a smaller step size. As NNs tend to become more inaccurate at the edges of the training data, we narrow the temperature range slightly. For the test set, the ranges change for the temperature from 25 °C to 130 °C, and for V_{ds} from 0.05 V to 0.8 V. Independent of the number of data points used in the training, the validation uses a temperature step size of 1 °C and a voltage step size of 0.01 V.

Table 2 Relative error of transistor characteristic quantiles depending on V_{ds}, V_{gs}, and temperature steps

V_{ds} steps	V_{gs} steps	Temp. steps	Data points	Q5 I_{off}	Q95 I_{off}	Q5 I_{on}	Q95 I_{on}	Q5 V_{th}	Q95 V_{th}	Q95 SS
80	12	6	5760	−5.02%	1.63%	−0.30%	0.51%	−0.50%	0.61%	4.11%
80	9	6	4320	−4.70%	2.07%	−0.41%	0.48%	−0.45%	0.75%	3.67%
80	7	6	3360	−4.92%	1.77%	−0.26%	0.46%	−0.06%	1.65%	4.35%
80	5	6	2400	−3.82%	1.86%	−0.38%	0.44%	−0.31%	2.14%	5.28%
17	17	6	1734	−5.43%	1.68%	−0.39%	0.48%	−0.73%	0.47%	4.30%
17	12	6	1224	−4.88%	1.67%	−0.22%	0.50%	−0.58%	0.49%	4.00%
17	9	6	918	−4.60%	1.77%	−0.43%	0.49%	−0.55%	0.76%	3.46%
12	17	6	1224	−6.62%	1.61%	−0.29%	0.62%	−0.84%	0.65%	5.11%
12	12	6	864	−6.91%	1.44%	−0.29%	0.52%	−0.72%	0.62%	5.92%
12	9	6	648	−5.62%	1.99%	−0.40%	0.56%	−0.63%	0.74%	4.56%

Highlight marks minimum data points satisfying the error limit of 5%

For each metric, the relative errors across the test set are calculated. Then, the error values are sorted so that a minimum error at the 5%-quantile (Q5) and a maximum error at the 95%-quantile (Q95) can be determined. This procedure is slightly different for SS as it can be measured at multiple points below the threshold voltage. We measure the SS relative error at multiple points and calculate the average of the absolutes, so that negative and positive relative errors do not cancel each other out. In consequence, there is only a Q95 error for SS. All metrics as well as the R^2 score are shown in Table 3.

For our evaluation, we define an error of $\leq$*5% at each 5%/95%-quantile to be acceptable.* However, please note that other values could have been chosen and internal experiments have shown that longer training times or more training data are sufficient for more stringent accuracy constraints.

All transistor characteristic results in Table 3 are created by using two separate neural networks with different scaling where the results are interpolated as described in Sect. 4.2. After 50 epochs, the R^2 score is already >0.99; however, Q5 I_{off} and Q95 SS still show an error of more than 11%, violating our tolerance. *This shows that the* R^2 *score of the NN is not sufficient to judge the accuracy of important transistor characteristics.* For the conventional FinFET device, the NN needs 400 epochs of training to reach the demanded precision.

4.7 Early Evaluation with Limited Data

To be able to reach an error $\leq$5% for all transistor characteristics, a certain amount of training data is needed. The required amount depends on the number of parameters and the number of measurement points for each parameter (i.e., in our case, sweeping steps in SPICE). Table 2 explores the relative error of each

Table 3 Comparison of R^2 score and transistor-specific characteristics depending on the number of trained epochs

Epochs	50	100	200	300	400
R^2 linear	0.99948	0.99968	0.99981	0.99989	0.99991
R^2 log	0.99666	0.99795	0.99896	0.99944	0.99968
Q5 I_{off}	−1.47%	−9.69%	−8.20%	−6.14%	−4.62%
Q95 I_{off}	3.01%	2.71%	2.62%	2.59%	2.93%
Q5 I_{on}	−1.93%	−1.68%	−1.28%	−1.14%	−0.96%
Q95 I_{on}	1.95%	1.80%	1.46%	1.30%	1.09%
Q5 V_{th}	−0.64%	−0.76%	−0.72%	−0.61%	−0.59%
Q95 V_{th}	2.12%	1.57%	1.17%	0.92%	0.67%
Q95 SS	1.25%	9.49%	7.78%	5.85%	4.45%

transistor characteristic depending on V_{ds} and V_{gs} steps. For the sake of space in this manuscript, a minimum of 6 temperature steps has been chosen as a suitable minimum based on preceding experiments. Table 2 shows that for a NN transistor model with temperature dependency, around 900 data points are sufficient to reach our tolerance. Looking at the step sizes, *we can say that around 17* $I_d - V_{gs}$ *curves with 9 data points each are enough to start developing an early model that satisfies our error constraint.* This number of points (measurements) is feasible to acquire even for early prototypes. With our selection of step sizes and data ranges, we are able to minimize the training effort to these 900 points.

4.8 *Modeling NC-FinFET with NN-Based Transistor Models*

To evaluate if our NN can also fit other, previously unseen transistors, we repeat the experiment in Sect. 4.6 using an NC-FinFET compact model [22] instead of BSIM-CMG. Comparing conventional FinFET and NC-FinFET, a stark difference in the behavior becomes apparent when looking at the transfer curves in Fig. 9. We can observe that compared to FinFET, NC-FinFET shows a clearly different, non-monotonic response in the V_d axis. This additional complexity is added by the negative DIBL effect caused by the ferroelectricity of the gate and suggests an increased difficulty in fitting the behavior. Table 4 shows the progress of learning NC-FinFET. When comparing these results with the FinFET results in Table 3, we can see that (especially in the first 200 epochs) almost all metrics show a worse error. Some stand out, e.g., Q5 I_{on}, with an error of −5.9% after 50 trained epochs (highlighted in Table 4). With an increasing number of training epochs, these errors are getting closer to the errors of the FinFET measurements. The desired precision of ≤5% relative error for each Q5/Q95 metric is reached after 400 epochs. *Eventually, our NN is able to apply to NC-FinFET in a similar quality as to FinFET.* This hints to the possibility that NN-based transistor models are indeed generic and thus that additional emerging technologies can be modeled with this approach.

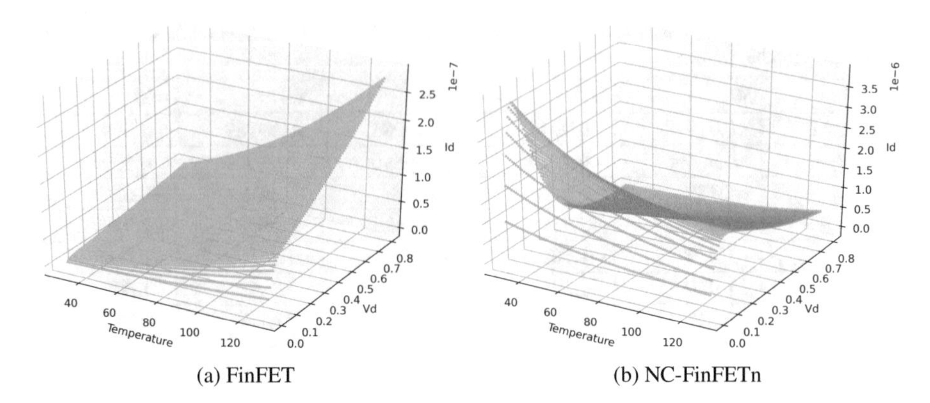

(a) FinFET (b) NC-FinFETn

Fig. 9 Comparison of FinFET and NC-FinFET transfer curves at $V_{gs} = 0.15\,\text{V}$. The negative DIBL effect in NC-FinFET completely changes the shape of the I–V curve

Table 4 Development of NN accuracy for NC-FinFET

Epochs	50	100	200	300	400
R^2 linear	0.99876	0.99956	0.99984	0.99988	0.99992
R^2 log	0.99552	0.99743	0.99879	0.99937	0.99960
Q5 I_{off}	−1.56%	−0.86%	−7.85%	−5.68%	−4.69%
Q95 I_{off}	4.83%	3.17%	3.22%	2.52%	2.67%
Q5 I_{on}	−5.90%	−4.35%	−2.31%	−1.86%	−1.43%
Q95 I_{on}	3.13%	1.70%	1.79%	1.40%	1.11%
Q5 V_{th}	−1.10%	−0.67%	−0.79%	−0.70%	−0.69%
Q95 V_{th}	3.95%	3.28%	1.94%	1.44%	1.18%
Q95 SS	2.11%	0.55%	7.22%	5.12%	4.44%

Compared to Table 3, NC-FinFET starts with higher errors but eventually converges to a similar precision of FinFET

5 Our Proposed Machine Learning-Based Approach

Analogously to the DTCO flow shown in Fig. 1, our approach consists of two major parts, one to be handled by the foundry, the other one by the designer. An overview is shown in Fig. 10: The first part is denoted by *Step 1* and includes the generation of sample libraries to provide training data, as well as the ML training itself. On the other side, there is *Step 2* which is cell library prediction and parameter optimization by the designer. The optional *Step 3* is comparatively small and its sole purpose is to make sure that the predicted performance is validated by a traditionally generated cell library. All details of Fig. 10 are explained in the following sections.

Our prototype implementation is mainly done in Python and C++. ML functionality is taken from Scikit-learn [25]. We also adopt their term *estimator* to refer to an instance of an ML algorithm. SciPy's *optimize* module [14] is used to minimize the objective function during parameter optimization. Established EDA tools by

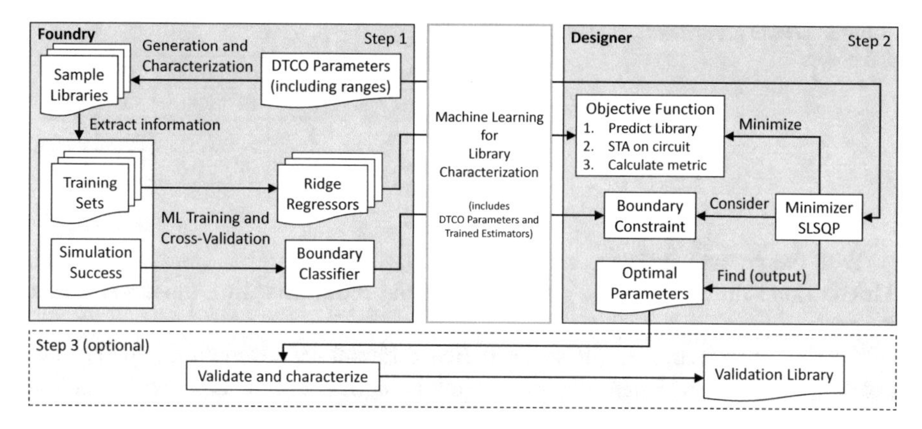

Fig. 10 Implementation of the DTCO process. Step 1: Data generation and training of the ML estimators by the foundry. Step 2: Design space exploration by the designer. Step 3: Traditional generation of cell library with optimized parameters for validation

Synopsys are used for cell library characterization, library compilation, and static timing analysis (STA).

5.1 *ML for Library Characterization*

The foundry part of our approach addresses all steps necessary to provide the designer with all the information they need for the DTCO process. This includes the generation of training data, as well as the training of estimators.

5.1.1 Generation of Training Data

The first step in our approach is the selection of *parameters* that should later be available for optimization, along with their value ranges. In general, a *parameter* could be any setting along the library characterization process. For details of the library characterization process, see our earlier work about reliability in standard cells [32, 33]. In this work, we chose to select T_{FE}, E_C, and P_R as parameters related to NC-FinFET, as well as V_{DD} which is part of the characterization settings. The boundaries of ferroelectric parameters shown in Table 5 are chosen by investigating NC-FinFET related papers [1, 2, 6, 12, 17, 18, 21, 23, 27, 29]. The value steps are not chosen equidistantly but lean towards the areas where more promising results are expected. By increasing the density of sample points in certain regions, we aim to reach a better accuracy in those areas.

Table 5 DTCO parameters and ranges

Name	Unit	Min				Max
E_C	MV/cm	100	200	400	700	1000
P_R	μC/cm^2	5	10	20	30	40
T_{FE}	nm	1	2	4	6	8
V_{DD}	V	0.3	0.4	0.5	0.6	0.8

With the parameters being selected, appropriate training data can be generated. This is done by characterizing multiple sample libraries within those parameter ranges.

As shown in Table 5, we selected five different values for each parameter, resulting in $5^4 = 625$ permutations. In addition, 250 combinations with random parameter values are selected. This leads to 875 sample libraries that are created to provide the training data. Characterizing this amount of libraries takes multiple weeks. From all libraries, 482 characterizations (55.2%) are finished without errors and are usable for training (90%), cross-validation (5-fold), and testing (10%).

5.1.2 Training of ML Estimators

For the library prediction, hundreds of estimators are created, one for each data field in the library. A data field can correspond to a single value (e.g., capacitance for a certain pin) or a 1D/2D-table (e.g., cell delays for a specific timing arc). Likewise, the previously generated sample libraries are split up by these fields and form training sets for the respective estimators.

Each of those estimators is a Ridge Regressor, performing linear regression with regularization. As well established in ML, the input parameters are extended with polynomial features so that polynomial functions can be fitted. This is done automatically for each estimator. The degree of the polynomial function and the regularization term are chosen by hyperparameter optimization.

Using polynomial functions to match the training sets has up- and downsides:

+ For most parameters, some linear or exponential correlation between input and output is expected. Thus, a polynomial function usually fits these parameters very well.
+ Compared to neural networks, a simple estimator fits quickly and well, even if the training set is rather small. This is helpful in our work as data generation is very expensive.
– In general, there might be transistor or cell models that cannot be sufficiently described with a polynomial function.

In addition to the regression estimators for library prediction, we create a classification estimator to learn parameter boundaries. Although the individual parameter ranges are manually picked and might be valid on their own, certain combinations can still lead to an overall invalid combination, i.e., a not properly functioning transistor or cell. This leads to a high number of failing characterizations during the generation of training data. For the inference, we want to ensure that predicted libraries for similar combinations are considered to be invalid as well. Otherwise, we risk to predict unrealistic libraries and draw incorrect conclusions. This classification estimator is a Support Vector Machine (SVM) that predicts the likelihood of a valid configuration, given a set of parameters. It is called *Boundary Classifier* in Fig. 10, and it is later used to guide the boundaries of exploration space during parameter optimization.

After all estimators are trained, all information can be bundled up and forwarded to the designer. As shown in the middle of Fig. 10, this collection consists of the selection of parameters, trained estimators for library prediction, as well as a trained classifier for boundary limitation. This replaces the single-cell library in the original DTCO flow.

5.2 Design Technology Co-Optimization for NC-FinFET

With the previously generated information, the designer is able to perform DTCO as outlined on the right side of Fig. 10.

Prediction of Cell Libraries On the designer's side, the trained estimators can be used to build complete cell libraries. Alongside the estimators which predict values for the individual data fields, a cell library template is used, containing all constant (i.e., parameter independent) information. In our prototype, a custom tool merges predicted values and the template to form a complete standardized library file, compatible with existing EDA tools.

In addition to the library file, the Boundary Classifier tells the designer how likely this set of parameters would result in a successful library characterization. If the probability is low, the predicted library is to be considered invalid.

Automatic Parameter Optimization Parameter optimization can be achieved through established optimization algorithms. All we need is a function that takes a set of parameters as input and the performance metric of our choice as the output. In our work, this function implements the following steps:

1. Predict the cell library for the given set of parameters.
2. Run STA on the designer's circuit using the predicted cell library.
3. Extract the critical path delay from the generated timing report.

In general, the metric could also consider any other available information, such as leakage, power consumption, or even input parameter values.

Although any optimization algorithm could be used, we chose Sequential Least Squares Programming (SLSQP) [15] in our approach. It allows the configuration of arbitrary constraint functions in addition to fixed, absolute parameter boundaries. This way, we can consider the output of the Boundary Classifier during optimization and ensure that we do not encounter invalid libraries while moving through parameter space.

6 Evaluation and Experimental Results

As proof of concept, we set our focus on delay related results only. Thus, we will consider timing and capacitance information in the library evaluation and critical path delay when looking at STA results. Nevertheless, the same concepts can be applied for power consumption, leakage, etc. as well.

To verify our work, we have access to data mimicking the foundry as well as the designer. Hence we have access to an NC-FinFET compact model [22] calibrated with industrial measurements [13] and 14 nm FinFET cell netlists [31] for cell library characterization. This allows us to fully implement and evaluate the proposed DTCO flow. To mitigate the long characterization run-times for training data, we work with a reduced set of 31 cells in our experiments. The selection of cells are taken from an adder gate-level netlist: `AND2_X1`, `AND2_X2`, `AND3_X2`, `AOI21_X1`, `AOI21_X2`, `AOI22_X1`, `BUF_X2`, `CLKBUF_X4`, `CLKBUF_X8`, `INV_X1`, `INV_X12`, `INV_X2`, `INV_X4`, `INV_X8`, `NAND2_X1`, `NAND2_X2`, `NAND3_X1`, `NAND3_X2`, `NAND4_X1`, `NOR2_X1`, `NOR2_X2`, `NOR3_X1`, `NOR3_X2`, `OAI21_X1`, `OAI21_X2`, `OAI22_X1`, `OR2_X1`, `OR2_X2`, `OR4_X1`, `XNOR2_X1`, `XOR2_X1`. All circuits have been synthesized with these cells using the baseline configuration.

6.1 *Accuracy of Cell Library Prediction*

To get an initial idea, we have a look at the individual estimator predictions that are performed to build a cell library. In our test set predictions, the estimators show an overall R^2 score of 99.02% and 97.00% for all data fields related to capacitance and timing, respectively. Figure 11 shows a histogram of the individual estimators, grouped by data type. The majority achieve an R^2 score of well above 95%. But there are also some worse performing estimators, especially for timing-related information. The few worst performers are annotated in the graph and reach an R^2 score of 74–83%. The estimator names represent the position of the data field in the hierarchical structure of the library file.

A look at the estimator's learning curves gives us an insight into the reason for their performance. For the vast majority of estimators, the learning curve looks very similar. We picked a representative curve in Fig. 12a. It shows that the estimator model (after selection of hyperparameters) is suited to fit the training set (red curve) and that the number of training samples is big enough for the model to generalize well (green curve). For most of the weaker estimators, the learning curve indicates that the selected estimator model is not perfectly fitting the training data (Fig. 12b,

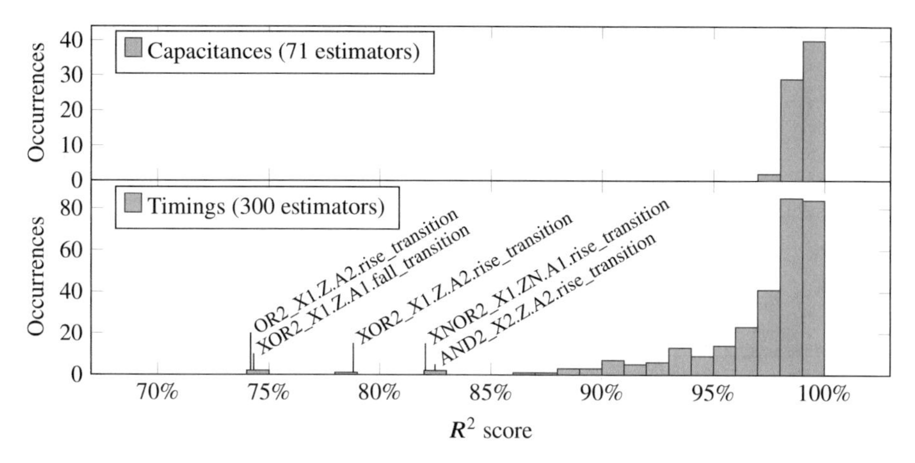

Fig. 11 Histogram of estimator accuracies. Each estimator is responsible to predict the values for a certain data field, i.e., combination of cell, pin(s), type, and timing arc. The few worst estimators and their appearances are annotated in the graph

c). In Fig. 12d, the plot suggests that more training samples could have improved that estimator slightly.

6.2 *Accuracy of Prediction on System Level*

After we observed the prediction accuracy for individual estimators, we want to see how the accuracy scales to the system level. Later, we use STA timing results to optimize our parameters in DTCO; thus, predicted libraries should perform well when being used in STA.

To evaluate system-level accuracy, we generate STA timing reports from predicted libraries, as well as from validation libraries build with the same parameters, and compare the path delays. With multiple paths per timing report, we can calculate an R^2 score the same way it is done for estimators. To make sure that our observations are general enough, we evaluate the STA performance on 11 different circuits. In Fig. 13, 10 path delays are reported for each set of parameters in the test set and for each circuit. We can see that the prediction accuracy is similar for all circuits.

The parameter ranges we select in Table 5 are not uniformly distributed; thus, we expect to experience different accuracies for certain regions in the ML models. In Fig. 14, we look again at STA accuracies for different circuits but in dependence of the parameter V_{DD}. We can observe that the accuracy drops at 0.7 V, which is also a gap in our selection of parameter values. Also smaller circuits are more impacted, with adder8dw dropping below an R^2 score of 80%.

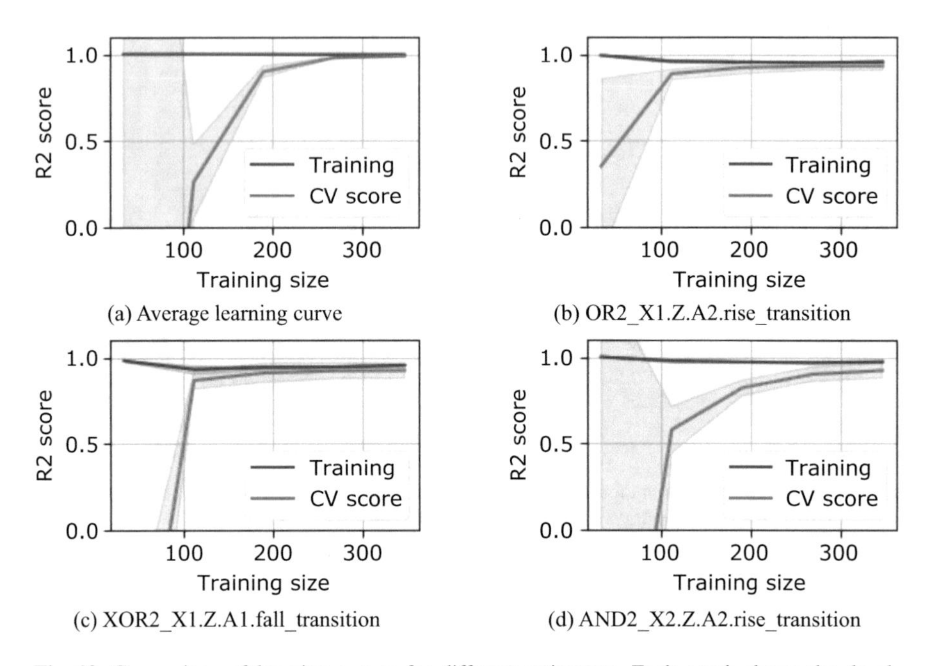

Fig. 12 Comparison of learning curves for different estimators. Each graph shows the development of R^2 score for the training set (red) and the cross-validation set (green) after training with a certain number of samples. Figure 12a shows NAND3_X1.ZN.A3.negative_unate.cell_fall as an average representative for the vast majority of learning curves among estimators. (**a**) Average learning curve. (**b**) OR2_X1.Z.A2.rise_transition. (**c**) XOR2_X1.Z.A1.fall_transition. (**d**) AND2_X2.Z.A2.rise_transition

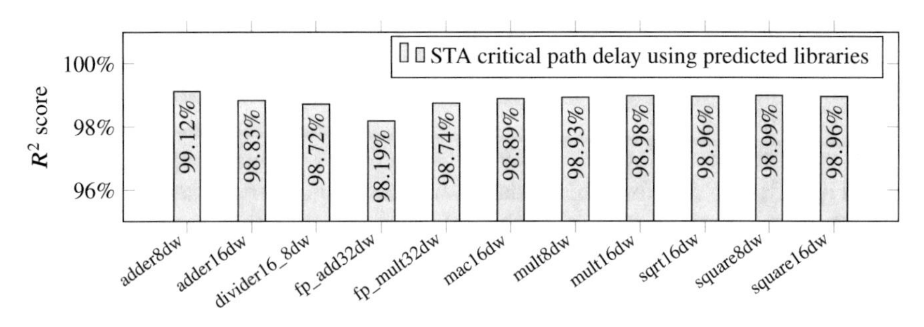

Fig. 13 Accuracy of critical path delay when using predicted libraries in STA. Test set libraries and prediction have been used here

In general, the prediction accuracy of libraries translates well to the system level. There is no major degradation in the R^2 score for different circuits. However, similar to the previous section, there are some regions where worse performance is to be expected.

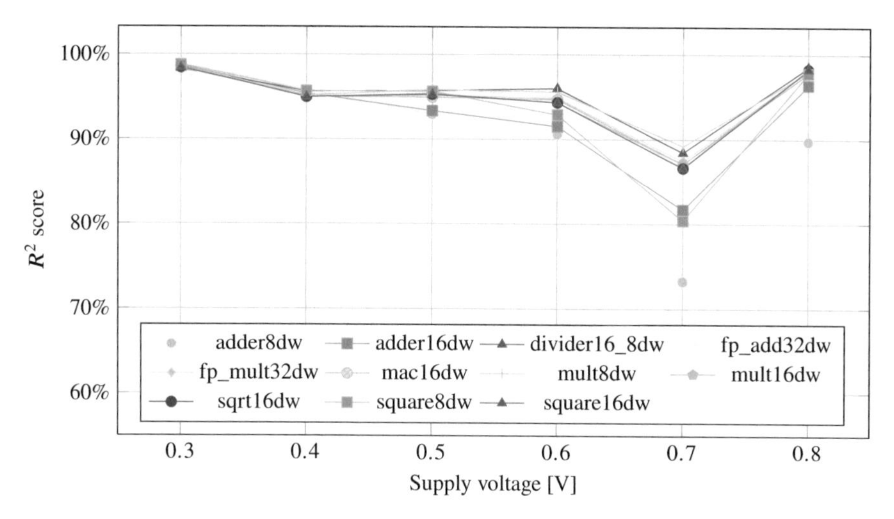

Fig. 14 STA accuracy as in Fig. 13 but for different voltages. The impact of skipping 0.7 V in the training parameters (see Table 5) is easy to observe

6.3 DTCO Parameter Optimization

Finally, we make use of our library prediction in the proposed DTCO workflow for automatic parameter optimization. We perform the optimization for two different scenarios and examine the results.

For NC-FinFET, minimizing the critical path delay means maximizing the amplification A_V in the transistor channel. This is achieved by matching the capacitance of the ferroelectric layer C_{fe} with the capacitance of the internal gate C_{int} as closely as possible [4].

$$A_V = \frac{\partial V_{\text{int}}}{\partial V_G} = \frac{|C_{\text{fe}}|}{|C_{\text{fe}}| - C_{\text{int}}} \tag{2}$$

The optimization algorithm will implicitly search for the best amplification by changing the ferroelectric material parameters E_C, P_R, and T_{FE} in order to minimize delay. But the gate capacitances are not only influenced by the NC-FinFET-related parameters but also by voltage. Because of that, we expect different optimal NC-FinFET parameters depending on V_{DD}. For our evaluation, we consider two scenarios with different upper limits for V_{DD}: 0.7 V and 0.5 V. As voltage has a big impact on the circuit delay, V_{DD} always reaches the upper boundary during optimization.

It is important to note that the ferroelectric parameters E_C, P_R, and T_{FE} form a Pareto-optimal space where the same performance can be achieved by multiple combinations. For example, a doubling of E_C can be negated by halving T_{FE} at the same time, resulting in the same delay. Due to this behavior, there are an arbitrary number of combinations for optimal performance, although not each combination would result in a valid configuration. This can also be observed directly in the transistor model implementation [22] and also in other papers investigating NC-FinFET parameters [26].

$$\text{Critical path delay} \propto \frac{P_R}{E_C \times T_{FE}} \tag{3}$$

To prevent arbitrary looking results and make comparison easier, we want to lay our focus only on one transistor parameter: T_{FE}. Therefore, we build a custom *objective function* for the optimization algorithm that taxes the change in P_R and E_C compared to the baseline configuration [13]. So on top of the delay function that we seek to minimize, we add penalties for two out of three parameters. As a result, the optimization algorithm will try to express the optimal parameters by changing T_{FE} rather than E_C or P_R. Looking back at Fig. 2, our *performance metric* of the optimization is implemented through this custom objective function.

For a set of parameters x, the custom objective function for minimization is

$$Objective(x) = Delay(x) \times \left(1 + \sum Penalty_p(x)\right) \tag{4}$$

with a penalty for each parameter $p \in \{P_R, E_C\}$ defined as

$$Penalty_p(x) = \alpha \times \left(\frac{Target_p - x_p}{Max_p - Min_p}\right)^2 \tag{5}$$

where $Target_p$ is the value to match (the calibrated model value), and α controls how strongly the target value is enforced. The parameter targets are given as $Target_{P_R} = 14.83$ and $Target_{E_C} = 125.59$ in the baseline reference.

With this objective function, we perform parameter optimization for each circuit individually. We set the upper limit for V_{DD} as a constraint function, alongside the Boundary Classifier constraint that ensures we end up with a result within a well-defined parameter space. To avoid getting stuck in local minima, each optimization is repeated 10 times with different, random starting points.

We perform the first optimization with an upper limit for $V_{DD} = 0.7\,\text{V}$. The outcome of the optimization for each circuit is shown in Table 6. The discovered, optimal parameters are listed in the left columns, whereas the right column shows the value of the objective function at that point. Since the algorithm tries to keep the penalties low, this value is close to the predicted critical path delay. As V_{DD} has a high impact on the circuit delay, we find V_{DD} at the upper boundary for all circuits. Due to the added penalties, E_C and P_R are close to the values of the calibrated baseline. The biggest difference can be observed in T_{FE} (highlighted in Table 6), changing roughly 0.8 nm between adder16dw and fp_add32dw. Figure 15 shows the critical path delay of the baseline configuration compared to the delay at the discovered optimal parameters. We can see that we are able to obtain a faster configuration for each circuit, compared to the baseline. However, we can also see

Table 6 Optimization results, iteratively searching from 10 random starting points for each circuit

	Explored optimal parameters				
Circuit	E_C	P_R	T_{FE}	V_{DD}	$Objective(x)$
adder8dw	133.3	14.03	4.548	0.7	0.05761
adder16dw	129.5	14.34	4.870	0.7	0.08224
divider16_8dw	128.9	14.37	4.408	0.7	0.49150
fp_add32dw	128.0	14.40	4.148	0.7	0.84740
fp_mult32dw	128.8	14.33	4.525	0.7	0.42450
mac16dw	129.1	14.34	4.493	0.7	0.26450
mult8dw	126.3	14.41	4.310	0.7	0.18030
mult16dw	126.9	14.45	4.388	0.7	0.29600
sqrt16dw	126.8	14.44	4.509	0.7	0.21340
square8dw	131.0	14.24	4.523	0.7	0.13280
square16dw	126.9	14.43	4.524	0.7	0.21340

Using custom objective function with $\alpha = 10$ and upper limit of $V_{DD} = 0.7\,\text{V}$

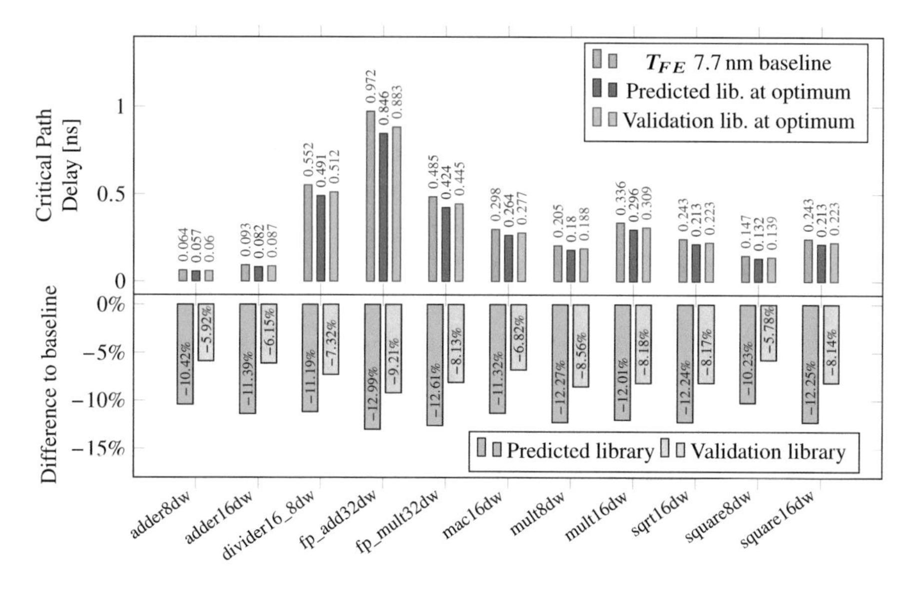

Fig. 15 Critical path delays for different circuits at 0.7 V. Using predicted libraries, the delay tends to be always too optimistic

that the predicted delay is always too optimistic when looking at the validation libraries for the same configuration.

The difference between prediction and validation can be explained by the behavior of the optimization algorithm. Always trying to find the minimal configuration has two effects. Firstly, from all variations in the accuracy, we always tend to see the *optimistic* side. A pessimistic outlier is unlikely to show up because it will always have a higher output in the objective function. Secondly, the biggest outliers are the

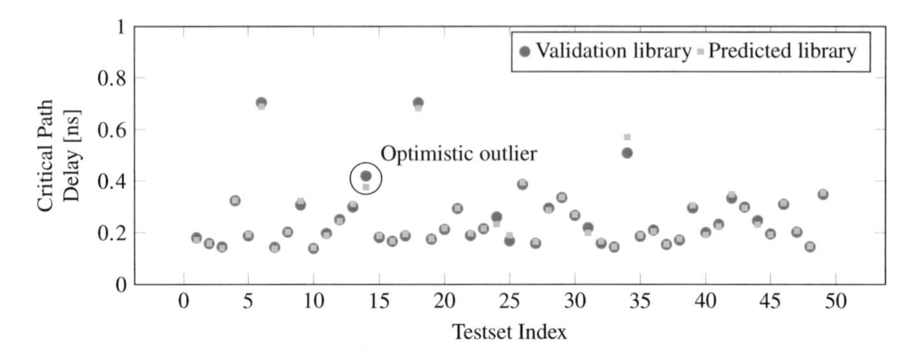

Fig. 16 Critical path delay on all configurations in the test set. Although the calculated R^2 score is high, the impact of outliers overshadow the result during automatic parameter optimization

most likely to be picked, if they are close enough to the optimum. Thus, we perceive a worse accuracy of the prediction compared to the previous section. The problem is visualized again in Fig. 16. Here, we see the critical path delay for all parameter configurations from the test set. Although the overall determination looks good at first glance, and the R^2 score is high (see Fig. 13), we can also see outliers that can become dominant in the parameter optimization.

6.4 Improvement in Performance

Technically, the proposed parameter optimization does not require ML to work. One could also generate libraries in the traditional way, for each point along the optimization process. However, ML is needed to make the approach feasible for extensive explorations.

For each iteration in the optimization algorithm, the gradient of the objective function is approximated by sampling the function with a small delta in each dimension. Thus, the total number of function evaluations for one optimization run is

$$N_{\text{evaluation}} \approx N_{\text{iteration}} \times (1 + N_{\text{parameter}}) \tag{6}$$

In our evaluation scenarios, we observed an average iteration count of 17.82 per optimization run. With 4 parameters to explore and 10 runs per circuit, we predicted roughly 900 libraries per circuit. (Note that although the libraries are not circuit-specific, the search path of the optimization algorithm is different, thus more libraries need to be generated.)

Looking at the 875 sample libraries that have been generated for training, the proposed approach pays off right after the first optimization. The generation of sample libraries took about 2 weeks on a 6th-generation Intel Core i7 with 32 GB of memory. For the experiments demonstrated in this work, we predicted roughly 20,000 libraries. The prediction of a single library is done in a split second; however, library compilation and STA can take up to a minute (depending on the circuit) which makes one optimization take about a few hours for all circuits.

For many experiments and more thorough investigations, the proposed approach would become infeasible very quickly, without the use of machine learning.

7 Conclusion

In this work, we presented two ML approaches. The first transistor-level approach that is able to learn FinFET transfer curves with high accuracy specific to selected transistor characteristics. Without any changes to the NN, it is also able to learn emerging NC-FinFET, although extra complexity is added due to the negative DIBL effect. This hints at the chance that more emerging technologies can be estimated with ML techniques, serving as an intermediate solution until sophisticated compact models are developed.

An ML solution can not only be developed in shorter time frames than the traditional compact model, it is also easier to share with others. Due to the black-box characteristics of the NN, the extraction of technology details is impractical or even impossible. With these benefits at hand, ML provides a great opportunity to speed up technology development, achieving faster time-to-market, and increase customer acceptance due to easier access.

Additionally, our second cell-level approach for cell library generation enables thorough design space exploration with increased accessibility for a wider audience of circuit designers.

Evaluating the timing-related performance, the ML approach reaches an R^2 score of about 98% for individual library values as well as on system level. In the DTCO flow, the perceived accuracy drops as outliers are more likely to be interpreted as good configurations. Nevertheless, our approach was able to find a set of parameters resulting in a faster configuration in all tests.

Acknowledgements We want to thank Yogesh S. Chauhan for the NCFET Model and Jannik Prinz for the implementation of the ML transistor modeling.

References

1. Agarwal, H., Kushwaha, P., Duarte, J.P., Lin, Y.K., Sachid, A.B., Chang, H.L., Salahuddin, S., Hu, C.: Designing 0.5 v 5-nm hp and 0.23 v 5-nm lp nc-finfets with improved i_{OFF} sensitivity in presence of parasitic capacitance. IEEE Trans. Electron Devices **65**(3), 1211–1216 (2018)
2. Agarwal, H., Kushwaha, P., Lin, Y.K., Kao, M.Y., Liao, Y.H., Duarte, J.P., Salahuddin, S., Hu, C.: Ncfet design considering maximum interface electric field. IEEE Electron Device Lett. **39**(8), 1254–1257 (2018)

3. Amrouch, H., Pahwa, G., Gaidhane, A.D., Henkel, J., Chauhan, Y.S.: Negative capacitance transistor to address the fundamental limitations in technology scaling: processor performance. IEEE Access **6**, 52754–52765 (2018)
4. Amrouch, H., Salamin, S., Pahwa, G., Gaidhane, A.D., Henkel, J., Chauhan, Y.S.: Unveiling the impact of ir-drop on performance gain in ncfet-based processors. IEEE Trans. Electron Devices **66**(7), 3215–3223 (2019)
5. Amrouch, H., Pahwa, G., Gaidhane, A.D., Dabhi, C.K., Klemme, F., Prakash, O., Chauhan, Y.S.: Impact of variability on processor performance in negative capacitance finfet technology. IEEE Trans. Circuits Syst. I Regul. Pap. **67**(9), 3127–3137 (2020)
6. Bansal, M., Kaur, H.: Analysis of negative-capacitance germanium finfet with the presence of fixed trap charges. IEEE Trans. Electron Devices **66**(4), 1979–1984 (2019)
7. Buhrmester, V., Münch, D., Arens, M.: Analysis of explainers of black box deep neural networks for computer vision: A survey. arXiv preprint arXiv:1911.12116 (2019)
8. Ceyhan, A., Quijas, J., Jain, S., Liu, H.Y., Gifford, W., Chakravarty, S.: Machine learning-enhanced multi-dimensional co-optimization of sub-10nm technology node options. In: Proceedings of the 2019 IEEE International Electron Devices Meeting (IEDM), pp. 36–6. IEEE, New York (2019)
9. Chauhan, Y.S., Venugopalan, S., Karim, M.A., Khandelwal, S., Paydavosi, N., Thakur, P., Niknejad, A.M., Hu, C.C.: Bsim–industry standard compact mosfet models. In: 2012 Proceedings of the European Solid-State Device Research Conference (ESSDERC), pp. 46–49. IEEE, New York (2012)
10. Dunga, M.V., Lin, C.H., Niknejad, A.M., Hu, C.: BSIM-CMG: A compact model for multi-gate transistors. In: Proceedings of the FinFETs and Other Multi-Gate Transistors, pp. 113–153. Springer, Berlin (2008)
11. Ha, D., Yang, C., Lee, J., Lee, S., Lee, S., Seo, K.I., Oh, H., Hwang, E., Do, S., Park, S., et al.: Highly manufacturable 7nm FinFET technology featuring EUV lithography for low power and high performance applications. In: Proceedings of the 2017 Symposium on VLSI Technology, pp. T68–T69. IEEE, New York (2017)
12. Hoffmann, M., Pešić, M., Slesazeck, S., Schroeder, U., Mikolajick, T.: Modeling and design considerations for negative capacitance field-effect transistors. In: Proceedings of the 2017 Joint International EUROSOI Workshop and International Conference on Ultimate Integration on Silicon (EUROSOI-ULIS), pp. 1–4. IEEE, New York (2017)
13. Hoffmann, M., Fengler, F.P., Herzig, M., Mittmann, T., Max, B., Schroeder, U., Negrea, R., Lucian, P., Slesazeck, S., Mikolajick, T.: Unveiling the double-well energy landscape in a ferroelectric layer. Nature **565**(7740), 464–467 (2019)
14. Jones, E., Oliphant, T., Peterson, P., et al.: SciPy: open source scientific tools for Python (2001). http://www.scipy.org/
15. Kraft, D., Schnepper, K.: SLSQP—a nonlinear programming method with quadratic programming subproblems. DLR, Oberpfaffenhofen (1989)
16. Lamamra, K., Berrah, S.: Modeling of mosfet transistor by mlp neural networks. In: Proceedings of the International Conference on Electrical Engineering and Control Applications, pp. 407–415. Springer, Berlin (2016)
17. Li, X., Sampson, J., Khan, A., Ma, K., George, S., Aziz, A., Gupta, S.K., Salahuddin, S., Chang, M.F., Datta, S., et al.: Enabling energy-efficient nonvolatile computing with negative capacitance fet. IEEE Trans. Electron Devices **64**(8), 3452–3458 (2017)
18. Lin, Y.K., Agarwal, H., Kao, M.Y., Zhou, J., Liao, Y.H., Dasgupta, A., Kushwaha, P., Salahuddin, S., Hu, C.: Spacer engineering in negative capacitance finfets. IEEE Electron Device Letters **40**(6), 1009–1012 (2019)
19. Meijer, P.B.L.: Neural Network Applications in Device and Subcircuit Modelling for Circuit Simulation. Philips Electronics (1996)
20. Mishra, S., Amrouch, H., Joe, J., Dabhi, C.K., Thakor, K., Chauhan, Y.S., Henkel, J., Mahapatra, S.: A simulation study of nbti impact on 14-nm node finfet technology for logic applications: Device degradation to circuit-level interaction. IEEE Trans. Electron Devices **66**(1), 271–278 (2018)

21. Pahwa, G., Dutta, T., Agarwal, A., Chauhan, Y.S.: Designing energy efficient and hysteresis free negative capacitance finfet with negative dibl and 3.5 xi on using compact modeling approach. In: Proceedings of the ESSCIRC Conference 2016: 42nd European Solid-State Circuits Conference, pp. 49–54. IEEE, New York (2016)
22. Pahwa, G., Dutta, T., Agarwal, A., Khandelwal, S., Salahuddin, S., Hu, C., Chauhan, Y.S.: Analysis and compact modeling of negative capacitance transistor with high on-current and negative output differential resistance–part i: Model description. IEEE Trans. Electron Devices **63**(12), 4981–4985 (2016). doi:10.1109/TED.2016.2614432
23. Pahwa, G., Dutta, T., Agarwal, A., Chauhan, Y.S.: Physical insights on negative capacitance transistors in nonhysteresis and hysteresis regimes: Mfmis versus MFIS structures. IEEE Trans. Electron Devices **65**(3), 867–873 (2018)
24. Paydavosi, N., Venugopalan, S., Chauhan, Y.S., Duarte, J.P., Jandhyala, S., Niknejad, A.M., Hu, C.C.: BSIM–SPICE models enable FinFET and UTB IC designs. IEEE Access **1**, 201–215 (2013)
25. Pedregosa, F., Varoquaux, G., Gramfort, A., Michel, V., Thirion, B., Grisel, O., Blondel, M., Prettenhofer, P., Weiss, R., Dubourg, V., Vanderplas, J., Passos, A., Cournapeau, D., Brucher, M., Perrot, M., Duchesnay, E.: Scikit-learn: Machine learning in Python. J. Mach. Learn. Res. **12**, 2825–2830 (2011)
26. Pentapati, S., Perumal, R., Khandelwal, S., Khan, A.I., Lim, S.K.: Optimal ferroelectric parameters for negative capacitance field-effect transistors based on full-chip implementations–part ii: Scaling of the supply voltage. IEEE Trans. Electron Devices **67**(1), 371–376 (2019)
27. Saeidi, A., Jazaeri, F., Bellando, F., Stolichnov, I., Enz, C.C., Ionescu, A.M.: Negative capacitance field effect transistors; capacitance matching and non-hysteretic operation. In: Proceedings of the 2017 47th European Solid-State Device Research Conference (ESSDERC), pp. 78–81. IEEE, New York (2017)
28. Salahuddin, S., Datta, S.: Use of negative capacitance to provide voltage amplification for low power nanoscale devices. Nano Lett. **8**(2), 405–410 (2008)
29. Sharma, A., Roy, K.: Design space exploration of hysteresis-free HFZRO x-based negative capacitance fets. IEEE Electron Device Lett. **38**(8), 1165–1167 (2017)
30. She, Y.q., Zhang, L.j., Zheng, J.b., Zhang, A.l., Zhu, Y.p., Li, Y.z.: Standard cell library characterization of 28nm process based on machine learning. In: DEStech Transactions on Computer Science and Engineering (CST) (2017)
31. Silvaco, Inc: Silvaco and si2 release unique, free 15nm open-source digital cell library (2019). https://www.silvaco.com/news/pressreleases/2019_05_30_01.html
32. van Santen, V.M., Amrouch, H., Henkel, J.: New worst-case timing for standard cells under aging effects. IEEE Trans. Device Mater. Reliab. **19**(1), 149–158 (2019)
33. van Santen, V.M., Amrouch, H., Henkel, J.: Modeling and mitigating time-dependent variability from the physical level to the circuit level. IEEE Transactions on Circuits and Systems I: Regular Papers. **66**(7), 2671–2684 (2019)
34. van Santen, V.M., et al.: Massively parallel circuit setup in GPU-SPICE. IEEE Trans. Comput. (2020)
35. Venugopalan, S., Paydavosi, N., Duarte, J., Lu, D., Khandelwal, S., Lin, C.H., Dunga, M., Yao, S., Niknejad, A., Hu, C.: Bsim-cmg 110 (2016). http://bsim.berkeley.edu/models/bsimcmg/
36. Wu, S.Y., Lin, C., Chiang, M., Liaw, J., Cheng, J., Yang, S., Tsai, C., Chen, P., Miyashita, T., Chang, C., et al.: A 7nm cmos platform technology featuring 4 th generation finfet transistors with a 0.027 um 2 high density 6-t sram cell for mobile soc applications. In: Proceedings of the 2016 IEEE International Electron Devices Meeting (IEDM), pp. 2–6. IEEE, New York (2016)
37. Zhang, L., Chan, M.: Artificial neural network design for compact modeling of generic transistors. J. Comput. Electron. **16**(3), 825–832 (2017)
38. Zhang, Z., Wang, R., Chen, C., Huang, Q., Wang, Y., Hu, C., Wu, D., Wang, J., Huang, R.: New-generation design-technology co-optimization (DTCO): Machine-learning assisted modeling framework. In: Proceedings of the 2019 Silicon Nanoelectronics Workshop (SNW), pp. 1–2. IEEE, New York (2019)

Fast Optimal Synthesis of Symmetric Index Generation Functions

Bernd Steinbach and Christian Posthoff

1 Introduction

A very basic task consists in finding an algorithm that can be used to solve a well-specified problem. This task has been solved in [1] for the problem to find a minimal circuit for the linear decomposition of symmetric index generation functions $S_1^n(\mathbf{x})$. This method uses a dynamic programming approach with the target of minimal general index generation functions. Solutions of $p_{\min}$ outputs y_j have been found for modules L of symmetric index generation functions $S_1^n(\mathbf{x})$ with $n = 10$ inputs x_i and gates with $1 \le t \le 5$ inputs. The computation of $p_{\min}$ have been aborted for $n \ge 10$ and $t \ge 2$, because the computation time was too long.

Knowing an algorithm for the given problem, a subsequent task consists in finding improved algorithms that solves the same problem faster and facilitates to solve larger problems of the same type. This subsequent task has been solved already twice; first in [2] and thereafter in [3]. These repeated efforts to improve the solution confirm the importance of the explored problem.

The branch-and-bound approach targeting only on symmetric index generation functions suggested in [2] shorten the time to compute the known solutions and found solutions $p_{\min}$ for larger problems up to $n = 30$ and $t = 5$. These problems have been solved several orders of magnitudes faster by an approach that utilizes symmetric properties of ZDDs in a dynamic programming algorithm suggested in

B. Steinbach (✉)
Institute of Computer Science, Freiberg University of Mining and Technology, Freiberg, Germany
e-mail: steinb@informatik.tu-freiberg.de

C. Posthoff
Department of Computing and Information Technology, The University of the West Indies, St. Augustine, Trinidad & Tobago
e-mail: christian@posthoff.de

R. Drechsler, D. Große (eds.), *Recent Findings in Boolean Techniques*,
https://doi.org/10.1007/978-3-030-68071-8_3

Table 1 Computation time of know methods for $S_1^n(\mathbf{x})$ (data taken from [3])

			Computation time in seconds for the solutions in		
Inputs n	Gate-inputs t	Outputs $p_{\min}$	[1] (2018)	[2] (2019)	[3] (2020)
10	1	9	$* < 0.01$	$* < 0.01$	$* < 0.01$
10	2	6	0.10	$* < 0.01$	$* < 0.01$
10	3	5	0.18	$* < 0.01$	$* < 0.01$
10	4	4	$* < 0.01$	$* < 0.01$	$* < 0.01$
10	5	4	$* < 0.01$	$* < 0.01$	$* < 0.01$
20	1	19	$* < 0.01$	$* < 0.01$	$* < 0.01$
20	2	13	†	0.03	$* < 0.01$
20	3	10	†	0.53	$* < 0.01$
20	4	8	†	1.92	$* < 0.01$
20	5	7	†	2.78	$* < 0.01$
30	1	29	$* < 0.01$	$* < 0.01$	$* < 0.01$
30	2	20	†	0.26	$* < 0.01$
30	3	15	†	9.19	$* < 0.01$
30	4	12	†	127.85	0.01
30	5	10	†	769.06	0.01

$* < 0.01$ means that the computation time is less than 0.01 s.
† means that the computation has been aborted since it was too long.

[3]. Table 1 summarizes the known results for such small problems; the data shown in this table have been taken from [3].

Problems up to $n = 80$ and $t = 5$ could be solved by the strongly improved approach of [3]; however, the needed computation time reveals that the computation time of this algorithm also increases exponentially with both the number n of inputs $\mathbf{x}$ and the number t of inputs of the gates of the module L.

Table 2 summarizes the known results for larger problems; the data shown in this table have been also taken from [3].

A faster heuristic approach is also provided in [3]; this approach finds in some case the minimal number $p_{\min}$ of outputs $\mathbf{y}$; but in some cases, up to six additional outputs are computed. We focus in this chapter to exact minimal results of $p_{\min}$ and skip therefore any heuristic approach.

The challenge of this chapter is the improvement of the approaches that are already improved two times; that means, we try to solve the same problems faster and try to find solutions $p_{\min}$ for larger values n of inputs $\mathbf{x}$ and larger values t of inputs of the gates used in the module L.

The rest of this chapter is structured as follows. Section 2 provides the used definitions and a fundamental Lemma. Section 3 utilizes the theorem of the orthogonality to reduce the cost of the circuit by substitutions of the used gates. Both the task to solve and our used approach are specified in Sect. 4. The core of this chapter is a detailed analysis of the reverse task to solve in Sect. 5. The results found in Sect. 5 are used in Sect. 6 to specify algorithms that compute either

Table 2 The so far best method for $S_1^n(\mathbf{x})$ (data taken from [3])

Inputs n	Gate-inputs t	Outputs $p_{\min}$	Computation time in seconds for the Solutions in [3] (2020)
40	1	39	$* < 0.01$
40	2	26	$* < 0.01$
40	3	20	0.01
40	4	16	0.07
40	5	13	0.18
50	1	49	$* < 0.01$
50	2	33	0.01
50	3	25	0.03
50	4	20	0.24
50	5	17	1.01
60	1	59	$* < 0.01$
60	2	40	0.01
60	3	30	0.06
60	4	24	0.63
60	5	20	3.50
70	1	69	$* < 0.01$
70	2	46	0.02
70	3	35	0.13
70	4	28	1.39
70	5	23	9.67
80	1	79	$* < 0.01$
80	2	53	0.02
80	3	40	0.22
80	4	32	2.72
80	5	27	1731.25

$* < 0.01$ means that the computation time is less than 0.01 s.

$p_{\min} = h_{\min}(n, t)$ or $n_{\max} = r_{\max}(p, t)$. Experimental results are provided in Sect. 7 before we conclude the results and specify some future work in Sect. 8.

2 Preliminaries

To be compatible with [1, 3, 5], we provide analogous basic definitions.

Definition 1 (Index Generation Function, Registered Vectors, Indices, Weight) An *index generation function* $f(\mathbf{x})$ is a multiple-valued function, where $\mathbf{x}$ is a tuple of n binary variables $(x_1, x_2, \ldots, x_n)$, and k assignments of values to these binary variables map to $K = \{1, 2, \ldots, k\}$. Hence, the variables of f are binary-valued, while f is k-valued. There is a one-to-one relationship between the k assignments of values to $(x_1, x_2, \ldots, x_n)$ and K; the other assignments are not specified. The k assignments of values to $(x_1, x_2, \ldots, x_n)$ are called the *registered vectors*. K is called the set of *indices*. $k = |K|$ is called the *weight* of the index generation function f.

Definition 2 (Characteristic Function) The *characteristic function* χ of an index generation function $f(\mathbf{x})$ is the logic function: $\{0, 1\}^n \to \{0, 1\}$ defined as

$$\chi(\mathbf{x}) = \begin{cases} 1 & \text{if } f(\mathbf{x}) \in K \\ 0 & \text{otherwise} . \end{cases}$$

We focus in this chapter to a special subset of index generation function for which the main definitions are provided next.

Definition 3 (Symmetric Function) A logic function $S(\mathbf{x})$ that satisfies

$$S(x_1, x_2, \ldots, x_i, \ldots, x_j, \ldots x_n) = S(x_1, x_2, \ldots, x_j, \ldots, x_i, \ldots x_n)$$

$\forall x_i, x_j \in \mathbf{x}$ is called a *symmetric function*; hence, the number of assignment of values 1 to $(x_1, x_2, \ldots, x_n)$ decides about the value of this function.

Definition 4 (Elementary Symmetric Function) An *elementary symmetric function* $S^n_m(\mathbf{x})$ is a special case of a symmetric function $S(\mathbf{x})$. $S^n_m(\mathbf{x})$ depends on n variables $\mathbf{x} = (x_1, x_2, \ldots, x_n)$ and is equal to 1 when values 1 are assigned to m of these variables.

Definition 5 (Symmetric Index Generation Function) Let $\chi(x_1, x_2, \ldots, x_n)$ be a characteristic function of an index generation function f. When χ is symmetric then f is a *symmetric index generation function*.

The explored symmetric index generation functions, which are explored in [3], have been restricted twice:

1. Only index generation functions $\chi(x_1, x_2, \ldots, x_n)$ of elementary symmetric function $S^n_m(\mathbf{x})$ are considered.
2. Out of the elementary symmetric function $S^n_m(\mathbf{x})$, only $S^n_1(\mathbf{x})$ has been explored in detail.

For comparison we use the same restrictions in this chapter.

The optimal realization of symmetric index generation functions $S^n_1(\mathbf{x})$ is strongly related to binomial coefficients. A central role for this synthesis has:

Lemma 1 *All integer $k \geq 2$ satisfy:*

$$1 \cdot \binom{k}{1} + 2 \cdot \binom{k}{2} = k^2 . \tag{1}$$

Proof

$$1 \cdot \binom{k}{1} + 2 \cdot \binom{k}{2} = \frac{1 \cdot k!}{1! \cdot (k-1)!} + \frac{2 \cdot k!}{2! \cdot (k-2)!} = \frac{k!}{(k-1)!} + \frac{k!}{(k-2)!}$$
$$= k + (k-1) \cdot k = k \cdot (1 + k - 1) = k \cdot k = k^2 \,. \qquad \square$$

3 Linearity, Orthogonality, and Circuit Structures

The linear decomposition of an index generation function has been suggested in [6]. Figure 1 shows the general circuit structure of such a linear decomposition.

The following definition specifies a linear decomposition used in [3] to realize a circuit of a symmetric index generation function.

Definition 6 (Linear Decomposition) A *linear decomposition* of an index generation function $f(x_1, x_2, \ldots, x_n)$ realizes f using a function $g(y_1, y_2, \ldots, y_p)$ storing indices and linear functions y_i:

$$y_i(x_1, x_2, \ldots, x_n) = a_{i1}x_1 \oplus a_{i2}x_2 \oplus \cdots \oplus a_{in}x_n \,, \qquad (2)$$

where $a_{ij} \in \{0, 1\}$, $i \in \{1, 2, \ldots, p\}$, $j \in \{1, 2, \ldots, n\}$, and, for all registered vectors of the index generation function, the following holds:

$$f(x_1, x_2, \ldots, x_n) = g(y_1, y_2, \ldots, y_p) \,.$$

The module L can be realized using EXOR-gates, because the functions $y_i(\mathbf{x})$ satisfy Eq. (2). A $(2^p \times q)$-bit memory has been suggested in [3] as implementation of the module G.

Example 1 Here we show the very simple example of a symmetric index generation function that has similarly been provided in [3]. Table 3a depicts the 4-variable symmetric index generation function $S_1^4(\mathbf{x})$ with the weight four. This function can be decomposed into two linear functions (we differently selected the variables to be conform with the rest of this chapter):

$$y_1 = x_2 \oplus x_4 \,,$$
$$y_2 = x_3 \oplus x_4 \,.$$

The function $g(y_1, y_2)$ is shown in Table 3b.

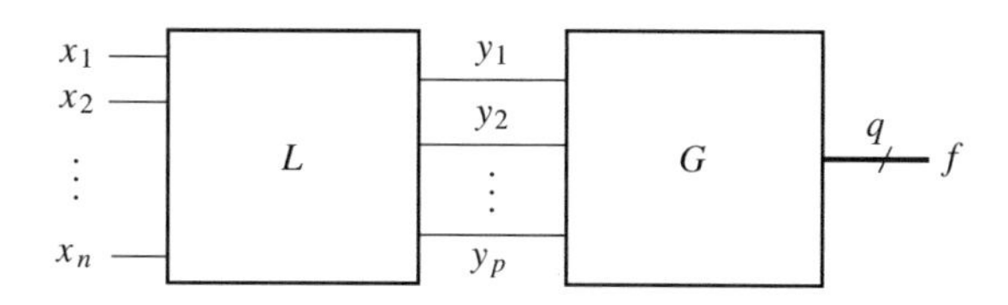

Fig. 1 Linear decomposition of index generation functions [6]

Table 3 Symmetric index generation function: (**a**) $S_1^4(\mathbf{x})$, (**b**) function $g(y_1, y_2)$ storing the indices of the linear decomposition of $S_1^4(\mathbf{x})$

a

Registered vectors				Indices of
x_1	x_2	x_3	x_4	$S_1^4(\mathbf{x})$
1	0	0	0	1
0	1	0	0	2
0	0	1	0	3
0	0	0	1	4

b

y_2	y_1	g
0	0	1
0	1	2
1	0	3
1	1	4

a

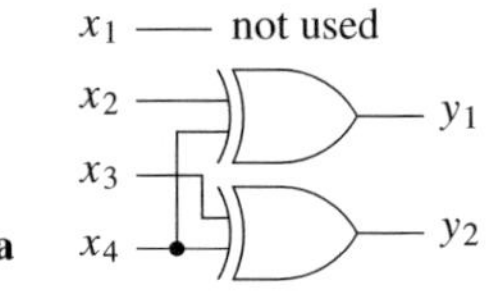

b

$x_1 \wedge x_2 = 0$
$x_1 \wedge x_3 = 0$
$x_1 \wedge x_4 = 0$
$x_2 \wedge x_3 = 0$
$x_2 \wedge x_4 = 0$
$x_3 \wedge x_4 = 0$

c

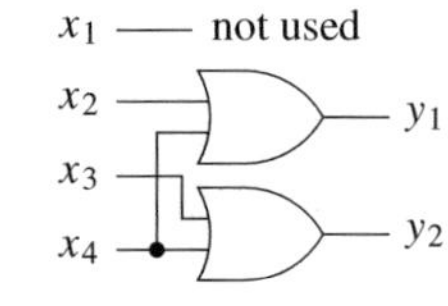

Fig. 2 Utilization of the orthogonality to realize the functions $y_1(\mathbf{x})$ and $y_2(\mathbf{x})$ of the symmetric function $S_1^4(\mathbf{x})$ of the module L: (**a**) circuit using EXOR-gates, (**b**) restrictions satisfied by $S_1^4(\mathbf{x})$, (**c**) circuit using OR-gates

It should be emphasized that in this function input values other than (1000), (0100), (0010), and (0001) are NOT assigned to any function values.

Each registered vectors of $S_1^n(\mathbf{x})$ contains only one single value 1; hence, these vectors satisfy the following:

Theorem 1 (Orthogonality [4]) *If the conjunctions C_i satisfy*

$$\forall i \neq j : \quad C_i \wedge C_j = 0$$

then

$$\bigvee_i C_i = \bigoplus_i C_i \, .$$

The linear decomposition used in Example 1 of $S_1^4(x_1, x_2, x_3, x_4)$ leads to the circuit structure of the module L shown in Fig. 2a. The variables $x_1, \ldots, x_4$ satisfy the condition of Theorem 1 (see Fig. 2b). Hence, the EXOR-gates in the circuit structure of symmetric index generation functions $S_1^4(\mathbf{x})$ can be replaced by OR-gates as shown in Fig. 2c. These substitutions reduce the cost of the circuit, but do not change the connection structure of the circuit.

The replacement of EXOR-gates by OR-gates is not restricted to circuits of $S_1^4(x_1, x_2, x_3, x_4)$, but can be used for all elementary symmetric functions $S_1^n(\mathbf{x})$ in the module L.

4 The Task to Solve and the Used Approach

The task to solve is determined by three integers that specify the module L of Fig. 1:

- n: the number of input variables x_i
- p: the number of output variables y_j
- t: the number of input variables of the gates in the module L

For easy comparison, we use the same names of variables as introduced in [3] where t has been defined by the less clear term *compound degree*.

The task to solve is:

$$p_{\min} = h_{\min}(n, t) \ . \tag{3}$$

That means, we are going to find the minimal number $p_{\min}$ of outputs y_j of the module L for elementary symmetric functions $S_1^n(\mathbf{x})$ depending on n input variables x_i, where each output y_j is created using an EXOR-gate (or better a simpler OR-gate) that has t inputs.

The efforts to solve this task using heuristics [3], exact solutions based on a dynamic programming approach [2], a branch-and-bound approach [1], or the much faster partition based approach [3] show that it is difficult to solve this task directly. Even the fastest approach of [3] has an exponential complexity.

Therefore, we are going to use a *two-step approach*:

1. Solve the *reverse task*:

$$n_{\max} = r_{\max}(p, t) \ ; \tag{4}$$

2. Determine $p_{\min}$ using the results of the first subtask:

$$p_{\min} = h'_{\min}(n, t) \ . \tag{5}$$

5 Analysis of the Properties of the Reverse Task

The reverse task consists in finding the maximal number of inputs x_i of the module L such that elementary symmetric functions $S_1^n(\mathbf{x})$ can be expressed by p outputs y_j using OR-gates (or EXOR-gates) of t inputs. It requires a deep analysis to find the function $n_{\max} = r_{\max}(p, t)$. In this section, we provide a sequence analysis step (structured by the following subsections) which result finally in a formula for the function $r_{\max}(p, t)$.

a	b	c	d
x_1 — not used	x_1 — not used	x_1 — not used	x_1 — not used
x_2 — y_1	x_2 — y_1	x_2 — y_1	x_2 — y_1
	x_3 — y_2	x_3 — y_2	x_3 — y_2
		x_4 — y_3	x_4 — y_3
			x_5 — y_4

Fig. 3 Circuit structures for the trivial case of $t = 1$: (**a**) $p = 1$, (**b**) $p = 2$, (**c**) $p = 3$, and (**d**) $p = 4$

5.1 *Trivial Solution for t = 1*

The value $t = 1$ determines that the gates of the module L have only a single input; hence, these gates can be omitted and simply replaced by wires.

The input pattern $x_1 = 1$ and $x_i = 0$ for all $i \neq 1$ can be represented by $y_j = 0$ for all $1 \leq j \leq i - 1$. The other input patterns have a single value 1 for x_i with $i > 1$; all other inputs are equal to 0. These patterns can be represented by $y_{j-i} = 1$ and values 0 for all other y_j. Hence, we get

$$n_{\max} = r_{\max}(p, 1) = p + 1 \,. \tag{6}$$

Figure 3 shows the circuits of modules L for $p = 1, \ldots, 4$ outputs y_i in which the OR-gates are replaced by wires due to the restriction of their inputs to $t = 1$.

5.2 *Smallest Optimal Circuits L for a Fixed Value of t*

Generally, the input pattern $\mathbf{x} = (x_1, x_2, x_3, \ldots, x_n) = (100\ldots0)$ can be mapped to $\mathbf{y} = (y_1, y_2, y_3, \ldots, y_p) = (000\ldots0)$; hence, no connection between the input x_1 and any input of the gates of the module L is needed for this case. Figure 4 shows this property by the input x_1 that is not connected with any gate of the circuit.

Figure 4 shows furthermore that the inputs x_2 to x_{p+1} can be connected to the first input of the p gates.

Consequently, $p \cdot (t - 1)$ inputs remain on the gates of the module L which can be used to uniquely encode further inputs x_i with $i > p + 1$. These inputs must be connected with inputs of at least two gates, because all output patterns $\mathbf{y}$ with exactly one value 1 have been already used to encode the inputs $x_2 = 1, \ldots, x_{p+1} = 1$. The number n of all inputs x_i is maximal when the inputs x_i with $i > p+1$ are connected with a minimal number of inputs of the gates in the module L, i.e., these inputs x_i are connected with exactly two inputs of two different gates.

Definition 7 (Smallest Optimal Number p_{so} of Gates in the Module L) A circuit of the module L with the *smallest optimal number* p_{so} of gates utilizes all $p_{so} \cdot t$

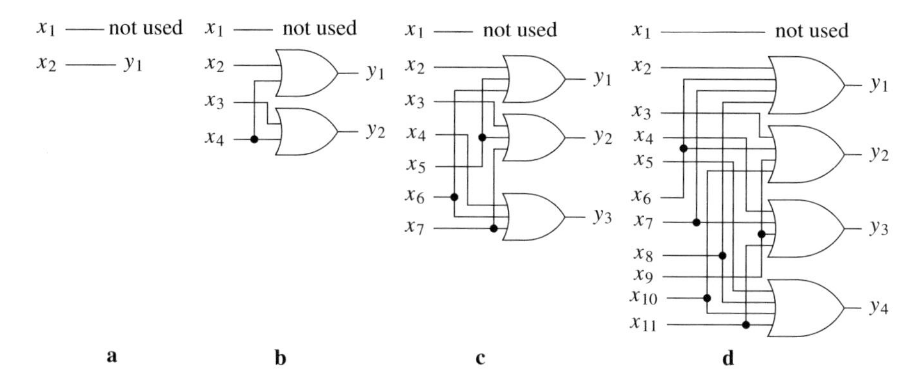

Fig. 4 Circuit structures of the smallest optimal number p_{so} of gates with t inputs: (**a**) $p_{so} = t = 1$, (**b**) $p_{so} = t = 2$, (**c**) $p_{so} = t = 3$, and (**d**) $p_{so} = t = 4$

inputs of the gates such that p_{so} inputs x_i are connected to one input of different gates (expressed by $1 \cdot \binom{p_{so}}{1}$) and all $\binom{p_{so}}{2}$ combinations of input pairs of the gates are used for additional inputs x_i; hence, we have:

$$p_{so} \cdot t = 1 \cdot \binom{p_{so}}{1} + 2 \cdot \binom{p_{so}}{2} . \tag{7}$$

Theorem 2 *The smallest optimal number p_{so} of gates in the module L is equal to the number of inputs of the used gates:*

$$p_{so} = t . \tag{8}$$

Proof Using Lemma 1, we get from Definition (7):

$$p_{so} \cdot t = p_{so}^2 , \qquad t = p_{so} , \qquad p_{so} = t . \qquad \square$$

The maximal number of inputs $n_{\max}$ of a circuit L with $p_{so} = t$ gates is the sum of:

- One input that is not connected with any gate (note: $1 = \binom{p_{so}}{0}$);
- $\binom{p_{so}}{1}$ inputs which are connected with only one input of p_{so} different gates
- $\binom{p_{so}}{2}$ inputs which are connected with two inputs of different gates such that all possible different combinations of gates are used

Hence, we get:

$$n_{\max} = r_{\max}(p_{so}, t = p_{so}) = 1 + \binom{p_{so}}{1} + \binom{p_{so}}{2} . \tag{9}$$

Table 4 Maximal number of inputs $n_{\max}$ for the smallest optimal number of gates p_{so} and the special case of $t = 1$

t	\	p	1	2	3	4	5	6	7	8	9	10
1			2	3	4	5	6	7	8	9	10	11
2				4								
3					7							
4						11						
5							16					
6								22				
7									29			
8										37		
9											46	
10												56

Table 4 shows the so far determined maximal numbers $n_{\max}$ of input variables x_i which can be represented using p_{so} gates with $t = p_{so}$ inputs as well as the special case for $t = 1$.

Figure 4 shows the circuits for the smallest optimal number of gates p_{so} and $t = 1, \ldots, 4$.

5.3 Regions of Restrictions

Several effects restrict the maximal number of inputs of the module L. There are three regions at all:

1. $1 \leq p \leq \log_2(2 \cdot t)$: the number $n_{\max}$ of inputs of the module L is restricted by the maximal number of different output patterns $\mathbf{y}$; in this region we have:

$$n_{\max} = r_{\max}(p, t) = 2^p \ . \tag{10}$$

2. $\log_2(2 \cdot t) < p < t$: at least one input x_i must be connected to more than two gates of the module L; hence, $p < p_{so}$; such cases occur for $t \geq 5$.
3. $p_{so} = t \leq p \leq \infty$: the number t of inputs of the gates of the module L restricts the number of utilized combinations to $\binom{p}{2}$ so that an optimal circuit can be built; for that reason, we restrict ourselves to this case usable for unrestricted large numbers of inputs $n_{\max}$ and realistic small numbers t of inputs of the gates of the module L.

The limit $p = \log_2(2 \cdot t)$ that separates the first two regions results from the binary encoding of maximal 2^p code-words of the length p. An arbitrary code-word $0 \leq i \leq 2^p$ and its mirrored code-word $0 \leq (2^p - i) \leq 2^p$ satisfy the properties that their conjunction is equal to the binary vector $\mathbf{0}$, and their disjunction is equal to the binary vector $\mathbf{1}$; hence, all such pairs of vectors require exactly p inputs of the gates of the module L. The number of inputs to encode the all 2^p different code-words is therefore

Table 5 Already determined values of the maximal number of inputs n_{max} and regions specified by the maximal number of connections of any input x_i with inputs of the gates of the module L

$t \setminus p$	1	2	3	4	5	6	7	8	9	10	11	12	13	14	15	16
1	2	3	4	5	6	7	8	9	10	11	12	13	14	15	16	17
2	2	4	$\leq$	$\leq$	$\leq$	$\leq$	$\leq$	$\leq$	$\leq$	$\leq$	$\leq$	$\leq$	$\leq$	$\leq$	$\leq$	$\leq$
3	2	4	7	$\leq$	$\leq$	$\leq$	$\leq$	$\leq$	$\leq$	$\leq$	$\leq$	$\leq$	$\leq$	$\leq$	$\leq$	$\leq$
4	2	4	8	11	$\leq$	$\leq$	$\leq$	$\leq$	$\leq$	$\leq$	$\leq$	$\leq$	$\leq$	$\leq$	$\leq$	$\leq$
5	2	4	8	$>$	16	$\leq$	$\leq$	$\leq$	$\leq$	$\leq$	$\leq$	$\leq$	$\leq$	$\leq$	$\leq$	$\leq$
6	2	4	8	$>$	$>$	22	$\leq$	$\leq$	$\leq$	$\leq$	$\leq$	$\leq$	$\leq$	$\leq$	$\leq$	$\leq$
7	2	4	8	$>$	$>$	$>$	29	$\leq$	$\leq$	$\leq$	$\leq$	$\leq$	$\leq$	$\leq$	$\leq$	$\leq$
8	2	4	8	16	$>$	$>$	$>$	37	$\leq$	$\leq$	$\leq$	$\leq$	$\leq$	$\leq$	$\leq$	$\leq$
9	2	4	8	16	$>$	$>$	$>$	$>$	46	$\leq$	$\leq$	$\leq$	$\leq$	$\leq$	$\leq$	$\leq$
10	2	4	8	16	$>$	$>$	$>$	$>$	$>$	56	$\leq$	$\leq$	$\leq$	$\leq$	$\leq$	$\leq$

$>$: region 2 determined by $\log_2(2 \cdot t) < p < t$; connections of at least one input x_i with more than two inputs (>2) of the gates of the module L are needed

$\leq$: region 3 determined by $p_{so} = t \leq p \leq \infty$; all inputs x_i are connected with maximal two inputs (≤ 2) of the gates of the module L

$$\frac{p \cdot 2^p}{2}$$

and the number of inputs of p gates is equal to $p \cdot t$. Hence, the explored limit is determined by

$$\begin{aligned}
\frac{p \cdot 2^p}{2} &= p \cdot t \\
\frac{2^p}{2} &= t \\
2^p &= 2 \cdot t \\
p &= \log_2(2 \cdot t)\ .
\end{aligned}$$

Table 5 shows the determined three regions by:

- Region 1: values 2^p on the left-hand side
- Region 2: symbols $>$ which indicate that at least one input x_i must be connected with more than two inputs of the gates of the module L
- Region 3: already determined values n_{max} and symbols $\leq$ which indicate that all inputs x_i are connected with not more than two inputs of the gates of the module L

In the remaining part of this chapter, we are going to explore only the region 3 specified by $p_{so} = t \leq p \leq \infty$.

5.4 Repeated Use of the Smallest Optimal Circuits

The numbers of inputs and outputs of the module L can be increased without any restrictions when the smallest optimal circuit is used several times. The maximal number of inputs $n_{\max}$ is:

$$n_{\max} = r_{\max}(p = k \cdot p_{so}, t = p_{so}) = 1 + k \cdot \left(\binom{p_{so}}{1} + \binom{p_{so}}{2} \right) , \qquad (11)$$

where k specifies how often the smallest optimal circuit has been used. It does not exist another circuit with more than $n_{\max}$ inputs x_i specified in (11), because all $k \cdot p_{so} \cdot t$ inputs of the $k \cdot p_{so}$ gates are used.

Figure 5 shows the circuits for $t = 1, \ldots, 4$ in which the circuit with the smallest optimal number of gates p_{so} has been used $k = 2$ times.

Table 6 summarizes the maximal numbers $n_{\max}$ determined by (6), (9), and (11).

It remains the task to determine the numbers $n_{\max}$ missing in Table 6 for $p > t$. In our detailed exploration, we noticed different results of this task for odd and even numbers of t; hence, we split our analysis for these two cases.

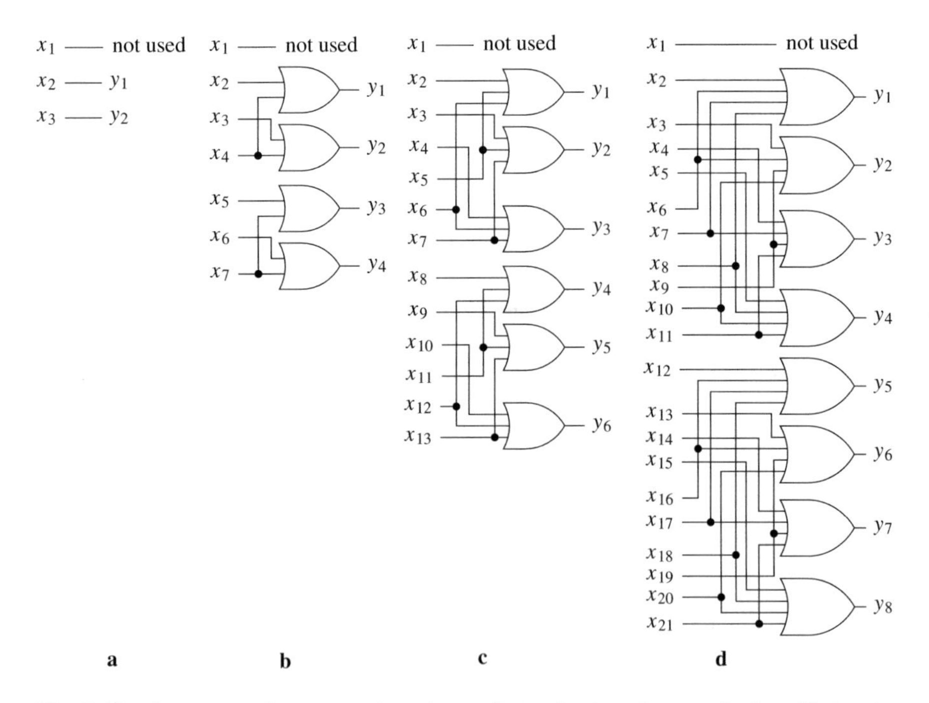

Fig. 5 Circuit structures that use two times the smallest optimal number p_{so} of gates with $t$inputs: **(a)** $p_{so} = t = 1$, **(b)** $p_{so} = t = 2$, **(c)** $p_{so} = t = 3$, and **(d)** $p_{so} = t = 4$

Table 6 Maximal number of inputs n_{max} for repeated k smallest optimal circuits

$t \setminus p$	1	2	3	4	5	6	7	8	9	10	11	12	13	14	15	16	17	18	19	20	21	22	23	24	25
1	2	3	4	5	6	7	8	9	10	11	12	13	14	15	16	17	18	19	20	21	22	23	24	25	26
2		4		7		10		13		16		19		22		25		28		31		34		37	
3			7			13			19			25			31			37			43			49	
4				11				21				31				41				51				61	
5					16					31					46					61					76
6						22						43						64						85	
7							29							57							85				
8								37								73								109	
9									46									91							
10										56										111					

5.5 *Missing Values n_{max} for Odd Values of t*

An odd integer $t > 1$ can be expressed by

$$t = 1 + l \cdot 2 ,$$

where l is a positive integer that is larger or equal to 1. Adding one gate with an odd number t of inputs to a smallest optimal circuit of the module L leads to t additional inputs on this gate which can be used as follows:

- One input of this gate is connected to an input x_i that is used only once
- l additional inputs x_i can be connected to two inputs of the gates such that different gates are used

Hence, n_{max} increases by $1 + l$ additional inputs x_i when p is increased by one gate with an odd number of inputs. The value $1+l$ of additional inputs x_i can alternatively be expressed by:

$$\begin{aligned} 1 + l &= 1 + \frac{t-1}{2} \\ &= \frac{t+1}{2} \end{aligned}$$

due to the definition of an odd integer given above.

The maximal number n_{max} of a module L consisting of $k \cdot p_{so} + d$ gates with an odd number t of inputs is:

$$\begin{aligned} n_{max} &= r_{max}(p = k \cdot p_{so} + d, t = p_{so}) \\ &= 1 + k \cdot \left(\binom{p_{so}}{1} + \binom{p_{so}}{2} \right) + d \cdot \frac{t+1}{2} , \end{aligned} \tag{12}$$

where $0 < d < t$.

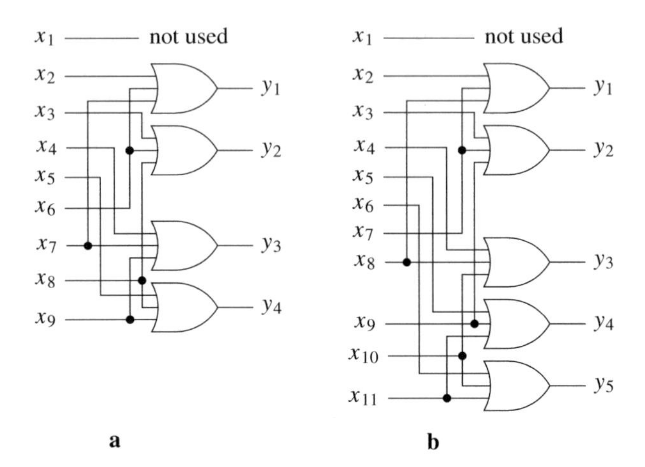

Fig. 6 Circuit structures for modules L of $p_{so} < p < 2 \cdot p_{so}$ gates with $t = 3$ inputs: (**a**) $p = 4$, $n_{\max} = 9$, and (**b**) $p = 5$, $n_{\max} = 11$

Figure 6 shows the circuit structure of the two circuits with $p_{so} < p < 2 \cdot p_{so}$ for gates with three inputs.

The missing values of $n_{\max}$ must be determined only for one interval between two repeated used smallest optimal circuits; the constructed circuits with $p \neq k \cdot p_{so}$ can be combined with k circuits of p_{so} inputs.

5.6 Missing Values $n_{\max}$ for Even Values of t

An even integer $t > 1$ can be expressed by

$$t = l \cdot 2 \,,$$

where l is a positive integer that is larger or equal to 1. We know from the previous analysis that odd numbers of t utilize all inputs of each additional gate; hence, for an even number of t and an odd number of p, one input of the additional gate remains unused.

Adding one gate with an even number t of inputs to a smallest optimal circuit (p_{so} is even) or to a larger circuit with an even number of p of the module L leads to t additional inputs on this gate which can be used as follows:

- One input of this gate is connected to an input x_i that is used only once
- $l - 1$ additional inputs x_i can be connected to two inputs of the gates such that different gates are used

Hence, $n_{\max}$ increases by $1 + (l - 1) = l$ additional inputs x_i when p is increased by one gate with an even number of inputs to an even number p.

The unused input of the gate in the circuit of an odd number $p > t$ and an even number t of inputs can be utilized when one more gate of t inputs is added; the available $t + 1$ inputs can be used as follows:

- One input of this gate is connected to an input x_i that is used only once
- l additional inputs x_i can be connected to two inputs of the gates such that different gates are used

Hence, $n_{\max}$ increases by $1 + (l) = l + 1$ additional inputs x_i when p is increased by one gate with an even number of inputs to an odd number p.

The maximal number $n_{\max}$ of a module L consisting of $p = k \cdot p_{so} + d$ gates with an even number t of inputs is:

$$\begin{aligned} n_{\max} &= r_{\max}(p = k \cdot p_{so} + d, t = p_{so}) \\ &= 1 + k \cdot \left(\binom{p_{so}}{1} + \binom{p_{so}}{2} \right) + \left\lfloor d \cdot \frac{t+1}{2} \right\rfloor , \end{aligned} \tag{13}$$

where $0 < d < t$. The pair of parentheses $\lfloor v \rfloor$ determines the largest integer that is smaller than or equal to the enclosed value v.

Figure 7 shows the circuit structure of the three circuits with $p_{so} < p < 2 \cdot p_{so}$ for gates with four inputs.

As in the case of odd numbers of t, the missing values of $n_{\max}$ must be determined only for one interval between two repeated used smallest optimal circuits with an

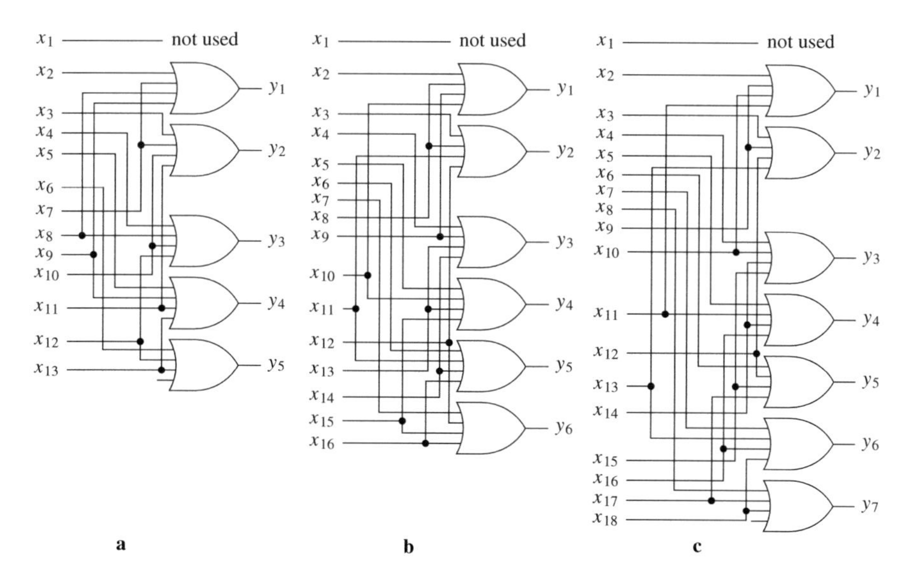

Fig. 7 Circuit structures for modules L of $p_{so} < p < 2 \cdot p_{so}$ gates with $t = 4$ inputs: (**a**) $p = 5$, $n_{\max} = 13$, (**b**) $p = 6$, $n_{\max} = 16$, and (**c**) $p = 7$, $n_{\max} = 18$

Table 7 Maximal number of inputs $n_{\max}$ for $p \geq t$

t	\	p	1	2	3	4	5	6	7	8	9	10	11	12	13	14	15	16	17	18	19	20	21	22	23	24	25
1			2	3	4	5	6	7	8	9	10	11	12	13	14	15	16	17	18	19	20	21	22	23	24	25	26
2				4	5	7	8	10	11	13	14	16	17	19	20	22	23	25	26	28	29	31	32	34	35	37	38
3					7	9	11	13	15	17	19	21	23	25	27	29	31	33	35	37	39	41	43	45	47	49	51
4						11	13	16	18	21	23	26	28	31	33	36	38	41	43	46	48	51	53	56	58	61	63
5							16	19	22	25	28	31	34	37	40	43	46	49	52	55	58	61	64	67	70	73	76
6								22	25	29	32	36	39	43	46	50	53	57	60	64	67	71	74	78	81	85	88
7									29	33	37	41	45	49	53	57	61	65	69	73	77	81	85	89	93	97	101
8										37	41	46	50	55	59	64	68	73	77	82	86	91	95	100	104	109	113
9											46	51	56	61	66	71	76	81	86	91	96	101	106	111	116	121	126
10												56	61	67	72	78	83	89	94	100	105	111	116	122	127	133	138

even number of t; the constructed circuits with $p \neq k \cdot p_{so}$ can be combined with k circuits of $t = p_{so}$ inputs.

5.7 *Summary of the Analysis*

Table 7 summarizes the results of the analysis of the maximal number of inputs $n_{\max}$ for $p \geq t$, $1 \leq t \leq 10$, and due to the available space $1 \leq p \leq 25$.

Equation (13) determines $n_{\max}$ for $p \geq t$ and even numbers t of inputs. This equation can also be used for odd numbers of t of inputs, because t is an odd number in (12), so that $\frac{t+1}{2}$ is an integer v and holds therefore $v = \lfloor v \rfloor$.

Using Eq. (8) of Theorem 2, we can simplify Eq. (13) furthermore:

$$
\begin{aligned}
n_{\max} &= 1 + k \cdot \left(\binom{p_{so}}{1} + \binom{p_{so}}{2} \right) + \left\lfloor d \cdot \frac{t+1}{2} \right\rfloor \\
&= 1 + k \cdot \left(\binom{t}{1} + \binom{t}{2} \right) + \left\lfloor d \cdot \frac{t+1}{2} \right\rfloor \\
&= 1 + k \cdot \left(\frac{t!}{1! \cdot (t-1)!} + \frac{t!}{2! \cdot (t-2)!} \right) + \left\lfloor d \cdot \frac{t+1}{2} \right\rfloor \\
&= 1 + k \cdot \left(t + \frac{(t-1) \cdot t}{2} \right) + \left\lfloor d \cdot \frac{t+1}{2} \right\rfloor \\
&= 1 + k \cdot \frac{2 \cdot t + (t-1) \cdot t}{2} + \left\lfloor d \cdot \frac{t+1}{2} \right\rfloor \\
&= 1 + k \cdot \frac{(t+1) \cdot t}{2} + \left\lfloor d \cdot \frac{t+1}{2} \right\rfloor .
\end{aligned}
$$

The product $(t+1) \cdot t$ is an even number for any value of t; hence, $\frac{(t+1) \cdot t}{2}$ is an integer so the whole term $k \cdot \frac{(t+1) \cdot t}{2}$ can be moved into the parentheses of the last term. Using furthermore $p = k \cdot t + d$, we get the universal equation for $p \geq t$:

$$n_{\max} = 1 + k \cdot t \cdot \frac{(t+1)}{2} + \left\lfloor d \cdot \frac{t+1}{2} \right\rfloor$$
$$= 1 + \left\lfloor (k \cdot t + d) \cdot \frac{t+1}{2} \right\rfloor$$
$$= 1 + \left\lfloor \frac{p \cdot (t+1)}{2} \right\rfloor . \quad (14)$$

The maximal number $n_{\max}$ of inputs x_i of the module L of a symmetric index generation function $S_1^n(\mathbf{x})$ can be computed in constant time using Eq. (14) for all positive integers $p \geq t$, where p is the number of gates of the module L and t is the number of inputs of the gates used in this module.

6 Algorithms to Compute $p_{\min}$ and $n_{\max}$

Table 7 shows that not all values n of input variables x_i belong to the set of maximal numbers $n_{\max}$ for given values of p gates with t inputs. The minimal number $p_{\min}$ for given values of n and t is the smallest value of p with $n \leq n_{\max}$. Algorithm 1 computes $p_{\min} = h'_{\min}(n, t)$ for $p_{\min} \geq t$ using a simple iteration over the values $n_{\max}$ determined by (14) starting with $p_{\min} = t$. No additional operation is needed to compute the smaller integer larger than or equal to the computed value in these parentheses, because the integer division by 2 directly computes this result.

Algorithm 1 $p_{\min} = h'_{\min}(n, t)$: Complete iteration

Input : n: number of input variables x_i of the module L for $S_1^n(\mathbf{x})$
Input : t: number of inputs of the gates of the module L for $S_1^n(\mathbf{x})$
Output : $p_{\min} \geq t$: minimal number of gates needed for the module L

1: $p_{\min} \leftarrow t$
2: **while** $n > ((((t+1) * p_{\min})/2) + 1)$ **do**
3: $\quad p_{\min} \leftarrow p_{\min} + 1$
4: **end while**

Algorithm 1 can be improved using the knowledge that uniquely $\left\lfloor \frac{t \cdot (t+1)}{2} \right\rfloor$ additional inputs x_i can be used for $p_{so} = t$ additional gates. This reduces the number of iterations to a value less than t. Algorithm 2 shows this faster approach.

Algorithms 1 and 2 can be extended such that $p_{\min}$ will be computed for

$$1 + \left\lfloor \frac{t \cdot (t+1)}{2} \right\rfloor \leq n \leq n_{\text{limit}} \quad \text{and} \quad 1 \leq t \leq t_{\text{limit}} ,$$

Algorithm 2 $p_{\min} = h'_{\min}(n, t)$: Reduced iteration

Input : n: number of input variables x_i of the module L for $S_1^n(\mathbf{x})$
Input : t: number of inputs of the gates of the module L for $S_1^n(\mathbf{x})$
Output : $p_{\min} \geq t$: minimal number of gates needed to realize the module L for $S_1^n(\mathbf{x})$

1: $p_{\min} \leftarrow (n-1)/(t*(t+1))/2)$
2: **while** $n > ((((t+1)*p_{\min})/2)+1)$ **do**
3: $\quad p_{\min} \leftarrow p_{\min}+1$
4: **end while**

where both n_{limit} and t_{limit} are arbitrarily chosen positive integers. In this case, the values of $n_{\max} = r_{\max}(p, t)$ must be computed several times. This repeated computation of the same values $n_{\max} = r_{\max}(p, t)$ can be avoided when these values are computed once, stored in a matrix, and reused several times for the computation of all required values $p_{\min}$. Algorithm 3 computes all values $n_{\max} = r_{\max}(p, t)$ for $1 \leq t \leq t_{\text{limit}}$ and $t \leq p \leq p_{\text{limit}}$, and stores the results in a matrix $n_{\max}[p, t]$.

Algorithm 3 $n_{\max} = r_{\max}(p, t)$: Reverse task

Input : p_{limit}: wanted maximal number of gates of the module L for $S_1^n(\mathbf{x})$
Input : t_{limit}: wanted maximal number of inputs of the gates of the module L for $S_1^n(\mathbf{x})$
Output : $n_{\max}$: maximal number of inputs x_i that control modules L for $S_1^n(\mathbf{x})$ in the ranges $1 \leq t \leq t_{\text{limit}}$ and $t \leq p \leq p_{\text{limit}}$

1: **for** $t \leftarrow 1$ **to** t_{limit} **do**
2: $\quad$ **for** $p \leftarrow t$ **to** p_{limit} **do**
3: $\quad\quad n_{\max}[p, t] \leftarrow 1 + ((p*(t+1))/2)$
4: $\quad$ **end for**
5: **end for**

Algorithm 4 computes all minimal values $p_{\min}$ of modules L for

$$1 \leq t \leq t_{\text{limit}} \quad \text{and} \quad 1 + \frac{t \cdot (t+1)}{2} \leq n \leq n_{\text{limit}} .$$

In the first two nested for-loop, the matrix $n_{\max}[p, t]$ is computed, p_{limit} is adjusted in line 2 to the required values of n. This matrix is used in the subsequent two nested for-loops, where Algorithm 2 evaluates only a subset of values of the matrix.

7 Experimental Results

We implemented all four algorithms of Sect. 6 and executed in the first experiment Algorithm 3 to solve the reverse task for modules L with $1 \leq t \leq t_{\text{limit}} = 100$ inputs of the gates and $t \leq p \leq p_{\text{limit}} = 1000$ gates. These 95,050 values $n_{\max}$ have been

Algorithm 4 $p_{\min} = h'_{\min}(n, t)$: Main task

Input : n_{limit}: wanted maximal number of inputs x_i of modules L for $S_1^n(\mathbf{x})$
Input : t_{limit}: wanted maximal number of inputs of the gates of modules L for $S_1^n(\mathbf{x})$
Output : $p_{\min}$: minimal number of gates of the module L for $S_1^n(\mathbf{x})$, $1 \leq t \leq t_{\text{limit}}$, and $1 + \left\lfloor \frac{t \cdot (t+1)}{2} \right\rfloor \leq n \leq n_{limit}$

```
1: for t ← 1 to t_limit do
2:     p_limit ← (n_limit − 1)/((t ∗ (t + 1))/2)
3:     for p ← t to p_limit do
4:         n_max[p, t] ← 1 + ((p ∗ (t + 1))/2)
5:     end for
6: end for
7: for t ← 1 to t_limit do
8:     for n ← n_max[t, t] to n_limit do
9:         p_min ← (n − 1)/(t ∗ (t + 1))/2)
10:        while n > n_max[p_min, t] do
11:            p_min ← p_min + 1
12:        end while
13:    end for
14: end for
```

computed within 0.05 ms on a PC Intel(R) Core(TM) i7-5960X CPU @ 3.00 GHz, using only 16 MB RAM, OS Windows 10 (64-bit), programming language: C++, and the execution environment Qt 5.14.1 with MinGW-64-bit. We executed all these values 10,000 times to measure this short times. Table 8 shows a small part of the values $n_{\max}$ computed by Algorithm 3.

Next we computed $p_{\min}(n = 80, t = 5) = 27$ using Algorithms 1 and 2. We executed each of these algorithms 1,000,000 times to measure a overall time difference; however, the needed time for these repeated computations was smaller than smallest measurable time interval of 1 ms. That means, we solved this task in less than 1 μs instead of almost half an hour needed for the solution of the same task in [3].

Finally, we implemented Algorithm 4 and used this C++ program to compute $p_{\min} = h'_{\min}(n, t)$. For larger values of t (inputs of the gates of the module L), less gates are needed to encode the active input x_i. Therefore, we solved this task for

$$1 \leq t \leq t_{\text{limit}} = 40 \quad \text{and} \quad 1 + \frac{t \cdot (t+1)}{2} \leq n \leq n_{\text{limit}} = 1000\,.$$

All these 28,520 minimal numbers $p_{\min}$ have been computed within 1 ms. Table 9 shows a small part of the values $p_{\min}$ computed by Algorithm 4.

Inside of Algorithm 4, we used Algorithm 2 that needs less sweeps in the while loop than Algorithm 1. The same results can be computed with Algorithm 4 in which line 9 (this is also the first line of Algorithm 2) is replaced by the simpler line: $p_{\min} \leftarrow t$, taken from the first line of Algorithm 1. We noticed in this experiment that both versions of Algorithm 4 require approximately the same time for the

Table 8 Subset of the computed solutions of the reverse task: $n_{\max} = r_{\max}(p, t)$

t	\	p	1	2	3	4	5	6	7	8	9	10	11	12	13	14	15	16	17	18	19	20	21	22	23	24	25	26	27	28	29	30
1			2	3	4	5	6	7	8	9	10	11	12	13	14	15	16	17	18	19	20	21	22	23	24	25	26	27	28	29	30	31
2				4	5	7	8	10	11	13	14	16	17	19	20	22	23	25	26	28	29	31	32	34	35	37	38	40	41	43	44	46
3					7	9	11	13	15	17	19	21	23	25	27	29	31	33	35	37	39	41	43	45	47	49	51	53	55	57	59	61
4						11	13	16	18	21	23	26	28	31	33	36	38	41	43	46	48	51	53	56	58	61	63	66	68	71	73	76
5							16	19	22	25	28	31	34	37	40	43	46	49	52	55	58	61	64	67	70	73	76	79	82	85	88	91
6								22	25	29	32	36	39	43	46	50	53	57	60	64	67	71	74	78	81	85	88	92	95	99	102	106
7									29	33	37	41	45	49	53	57	61	65	69	73	77	81	85	89	93	97	101	105	109	113	117	121
8										37	41	46	50	55	59	64	68	73	77	82	86	91	95	100	104	109	113	118	122	127	131	136
9											46	51	56	61	66	71	76	81	86	91	96	101	106	111	116	121	126	131	136	141	146	151
10												56	61	67	72	78	83	89	94	100	105	111	116	122	127	133	138	144	149	155	160	166
11													67	73	79	85	91	97	103	109	115	121	127	133	139	145	151	157	163	169	175	181
12														79	85	92	98	105	111	118	124	131	137	144	150	157	163	170	176	183	189	196
13															92	99	106	113	120	127	134	141	148	155	162	169	176	183	190	197	204	211
14																106	113	121	128	136	143	151	158	166	173	181	188	196	203	211	218	226
15																	121	129	137	145	153	161	169	177	185	193	201	209	217	225	233	241
16																		137	145	154	162	171	179	188	196	205	213	222	230	239	247	256
17																			154	163	172	181	190	199	208	217	226	235	244	253	262	271
18																				172	181	191	200	210	219	229	238	248	257	267	276	286
19																					191	201	211	221	231	241	251	261	271	281	291	301
20																						211	221	232	242	253	263	274	284	295	305	316
21																							232	243	254	265	276	287	298	309	320	331
22																								254	265	277	288	300	311	323	334	346
23																									277	289	301	313	325	337	349	361
24																										301	313	326	338	351	363	376
25																											326	339	352	365	378	391
26																												352	365	379	392	406
27																													379	393	407	421
28																														407	421	436
29																															436	451
30																																466
29																															436	451
30																																466
30																																466

Table 9 Subset of the computed solutions of the main task: $p_{\min} = h'_{\min}(n, t)$

n \ t	1	2	3	4	5	6	7	8	9	10	11	12	13	14	15
1															
2	1														
3	2														
4	3	2													
5	4	3													
6	5	4													
7	6	4	3												
8	7	5	4												
9	8	6	4												
10	9	6	5												
11	10	7	5	4											
12	11	8	6	5											
13	12	8	6	5											
14	13	9	7	6											
15	14	10	7	6											
16	15	10	8	6	5										
17	16	11	8	7	6										
18	17	12	9	7	6										
19	18	12	9	8	6										
20	19	13	10	8	7										
21	20	14	10	8	7										
22	21	14	11	9	7	6									
23	22	15	11	9	8	7									
24	23	16	12	10	8	7									
25	24	16	12	10	8	7									
26	25	17	13	10	9	8									
27	26	18	13	11	9	8									
28	27	18	14	11	9	8									
29	28	19	14	12	10	8	7								
30	29	20	15	12	10	9	8								
31	30	20	15	12	10	9	8								
32	31	21	16	13	11	9	8								
33	32	22	16	13	11	10	8								
34	33	22	17	14	11	10	9								
35	34	23	17	14	12	10	9								
36	35	24	18	14	12	10	9								
37	36	24	18	15	12	11	9	8							
38	37	25	19	15	13	11	10	9							
39	38	26	19	16	13	11	10	9							
40	39	26	20	16	13	12	10	9							
41	40	27	20	16	14	12	10	9							
42	41	28	21	17	14	12	11	10							
43	42	28	21	17	14	12	11	10							
44	43	29	22	18	15	13	11	10							
45	44	30	22	18	15	13	11	10							
46	45	30	23	18	15	13	12	10	9						
47	46	31	23	19	16	14	12	11	10						
48	47	32	24	19	16	14	12	11	10						
49	48	32	24	20	16	14	12	11	10						
50	49	33	25	20	17	14	13	11	10						
51	50	34	25	20	17	15	13	12	10						
52	51	34	26	21	17	15	13	12	11						
53	52	35	26	21	18	15	13	12	11						
54	53	36	27	22	18	16	14	12	11						
55	54	36	27	22	18	16	14	12	11						
56	55	37	28	22	19	16	14	13	11	10					
57	56	38	28	23	19	16	14	13	12	11					
58	57	38	29	23	19	17	15	13	12	11					
59	58	39	29	24	20	17	15	13	12	11					
60	59	40	30	24	20	17	15	14	12	11					
61	60	40	30	24	20	18	15	14	12	11					
62	61	41	31	25	21	18	16	14	13	12					
63	62	42	31	25	21	18	16	14	13	12					
64	63	42	32	26	21	18	16	14	13	12					
65	64	43	32	26	22	19	16	15	13	12					
66	65	44	33	26	22	19	17	15	13	12					
67	66	44	33	27	22	19	17	15	14	12	11				
68	67	45	34	27	23	20	17	15	14	13	12				
69	68	46	34	28	23	20	17	16	14	13	12				
70	69	46	35	28	23	20	18	16	14	13	12				
71	70	47	35	28	24	20	18	16	14	13	12				
72	71	48	36	29	24	21	18	16	15	13	12				
73	72	48	36	29	24	21	18	16	15	14	12				
74	73	49	37	30	25	21	19	17	15	14	13				
75	74	50	37	30	25	22	19	17	15	14	13				
76	75	50	38	30	25	22	19	17	15	14	13				
77	76	51	38	31	26	22	19	17	16	14	13				
78	77	52	39	31	26	22	20	18	16	14	13				
79	78	52	39	32	26	23	20	18	16	15	13	12			
80	79	53	40	32	27	23	20	18	16	15	14	13			
81	80	54	40	32	27	23	20	18	16	15	14	13			
82	81	54	41	33	27	24	21	18	17	15	14	13			
83	82	55	41	33	28	24	21	19	17	15	14	13			
84	83	56	42	34	28	24	21	19	17	16	14	13			
85	84	56	42	34	28	24	21	19	17	16	14	13			
86	85	57	43	34	29	25	22	19	17	16	15	14			
87	86	58	43	35	29	25	22	20	18	16	15	14			
88	87	58	44	35	29	25	22	20	18	16	15	14			
89	88	59	44	36	30	26	22	20	18	16	15	14			
90	89	60	45	36	30	26	23	20	18	17	15	14			

n \ t	1	2	3	4	5	6	7	8	9	10	11	12	13	14	15
91	90	60	45	36	30	26	23	20	18	17	15	14			
92	91	61	46	37	31	26	23	21	19	17	16	14	13		
93	92	62	46	37	31	27	23	21	19	17	16	15	14		
94	93	62	47	38	31	27	24	21	19	17	16	15	14		
95	94	63	47	38	32	27	24	21	19	18	16	15	14		
96	95	64	48	38	32	28	24	22	19	18	16	15	14		
97	96	64	48	39	32	28	24	22	20	18	16	15	14		
98	97	65	49	39	33	28	25	22	20	18	17	15	14		
99	98	66	49	40	33	28	25	22	20	18	17	16	14		
100	99	66	50	40	33	29	25	22	20	18	17	16	15		
101	100	67	50	40	34	29	25	23	20	19	17	16	15		
102	101	68	51	41	34	29	26	23	21	19	17	16	15		
103	102	68	51	41	34	30	26	23	21	19	17	16	15		
104	103	69	52	42	35	30	26	23	21	19	18	16	15		
105	104	70	52	42	35	30	26	24	21	19	18	16	15		
106	105	70	53	42	35	30	27	24	21	20	18	17	15	14	
107	106	71	53	43	36	31	27	24	22	20	18	17	16	15	
108	107	72	54	43	36	31	27	24	22	20	18	17	16	15	
109	108	72	54	44	36	31	27	24	22	20	18	17	16	15	
110	109	73	55	44	37	32	28	25	22	20	19	17	16	15	
111	110	74	55	44	37	32	28	25	22	20	19	17	16	15	
112	111	74	56	45	37	32	28	25	23	21	19	18	16	15	
113	112	75	56	45	38	32	28	25	23	21	19	18	16	15	
114	113	76	57	46	38	33	29	26	23	21	19	18	17	16	
115	114	76	57	46	38	33	29	26	23	21	19	18	17	16	
116	115	77	58	46	39	33	29	26	23	21	20	18	17	16	
117	116	78	58	47	39	34	29	26	24	22	20	18	17	16	
118	117	78	59	47	39	34	30	26	24	22	20	18	17	16	
119	118	79	59	48	40	34	30	27	24	22	20	19	17	16	
120	119	80	60	48	40	34	30	27	24	22	20	19	17	16	
121	120	80	60	48	40	35	30	27	24	22	20	19	18	16	15
122	121	81	61	49	41	35	31	27	25	22	21	19	18	17	16
123	122	82	61	49	41	35	31	28	25	23	21	19	18	17	16
124	123	82	62	50	41	36	31	28	25	23	21	19	18	17	16
125	124	83	62	50	42	36	31	28	25	23	21	20	18	17	16
126	125	84	63	50	42	36	32	28	25	23	21	20	18	17	16
127	126	84	63	51	42	36	32	28	26	23	21	20	18	17	16
128	127	85	64	51	43	37	32	29	26	24	22	20	19	17	16
129	128	86	64	52	43	37	32	29	26	24	22	20	19	18	16
130	129	86	65	52	43	37	33	29	26	24	22	20	19	18	17
131	130	87	65	52	44	38	33	29	26	24	22	20	19	18	17
132	131	88	66	53	44	38	33	30	27	24	22	21	19	18	17
133	132	88	66	53	44	38	33	30	27	24	22	21	19	18	17
134	133	89	67	54	45	38	34	30	27	25	23	21	19	18	17
135	134	90	67	54	45	39	34	30	27	25	23	21	20	18	17
136	135	90	68	54	45	39	34	30	27	25	23	21	20	18	17
137	136	91	68	55	46	39	34	31	28	25	23	21	20	19	17
138	137	92	69	55	46	40	35	31	28	25	23	22	20	19	18
139	138	92	69	56	46	40	35	31	28	26	23	22	20	19	18
140	139	93	70	56	47	40	35	31	28	26	24	22	20	19	18
141	140	94	70	56	47	40	35	32	28	26	24	22	20	19	18
142	141	94	71	57	47	41	36	32	29	26	24	22	21	19	18
143	142	95	71	57	48	41	36	32	29	26	24	22	21	19	18
144	143	96	72	58	48	41	36	32	29	26	24	22	21	20	18
145	144	96	72	58	48	42	36	32	29	27	24	23	21	20	18
146	145	97	73	58	49	42	37	33	29	27	25	23	21	20	19
147	146	98	73	59	49	42	37	33	30	27	25	23	21	20	19
148	147	98	74	59	49	42	37	33	30	27	25	23	21	20	19
149	148	99	74	60	50	43	37	33	30	27	25	23	22	20	19
150	149	100	75	60	50	43	38	34	30	28	25	23	22	20	19
151	150	100	75	60	50	43	38	34	30	28	25	24	22	20	19
152	151	101	76	61	51	44	38	34	31	28	26	24	22	21	19
153	152	102	76	61	51	44	38	34	31	28	26	24	22	21	19
154	153	102	77	62	51	44	39	34	31	28	26	24	22	21	20
155	154	103	77	62	52	44	39	35	31	28	26	24	22	21	20
156	155	104	78	62	52	45	39	35	31	29	26	24	23	21	20
157	156	104	78	63	52	45	39	35	32	29	26	24	23	21	20
158	157	105	79	63	53	45	40	35	32	29	27	25	23	21	20
159	158	106	79	64	53	46	40	36	32	29	27	25	23	22	20
160	159	106	80	64	53	46	40	36	32	29	27	25	23	22	20
161	160	107	80	64	54	46	40	36	32	30	27	25	23	22	20
162	161	108	81	65	54	46	41	36	33	30	27	25	23	22	21
163	162	108	81	65	54	47	41	36	33	30	27	25	24	22	21
164	163	109	82	66	55	47	41	37	33	30	28	26	24	22	21
165	164	110	82	66	55	47	41	37	33	30	28	26	24	22	21
166	165	110	83	66	55	48	42	37	33	30	28	26	24	22	21
167	166	111	83	67	56	48	42	37	34	31	28	26	24	23	21
168	167	112	84	67	56	48	42	38	34	31	28	26	24	23	21
169	168	112	84	68	56	48	42	38	34	31	28	26	24	23	21
170	169	113	85	68	57	49	43	38	34	31	29	26	25	23	22
171	170	114	85	68	57	49	43	38	34	31	29	27	25	23	22
172	171	114	86	69	57	49	43	38	35	32	29	27	25	23	22
173	172	115	86	69	58	50	43	39	35	32	29	27	25	23	22
174	173	116	87	70	58	50	44	39	35	32	29	27	25	24	22
175	174	116	87	70	58	50	44	39	35	32	29	27	25	24	22
176	175	117	88	70	59	50	44	39	35	32	30	27	25	24	22
177	176	118	88	71	59	51	44	40	36	32	30	28	26	24	22
178	177	118	89	71	59	51	45	40	36	33	30	28	26	24	23
179	178	119	89	72	60	51	45	40	36	33	30	28	26	24	23
180	179	120	90	72	60	52	45	40	36	33	30	28	26	24	23

same values of n_{limit} and t_{limit}. Obviously, the time to compute the division to get a larger starting value of $p_{\min}$ is nearly equal to the saved number of increments and comparisons $n > n_{\max}[p_{\min}, t]$ in the while loop.

8 Conclusion and Future Work

We solved in this chapter the task to find the minimal number of gates $p_{\min}$ of the modules L with n inputs x_i and t inputs of the used gates of the explored module L for $n \geq 1+\frac{t\cdot(t+1)}{2}$, where L realizes the symmetric index generation function $S_1^n(\mathbf{x})$. As result of a comprehensive analysis according to [7], we found Eq. (14) that solves the reverse task $n_{\max} = r_{\max}(p, t)$ in *constant time*. Using this intermediate result, we provided Algorithms 1 and 2 which solve the basic task $p_{\min} = h'_{\min}(n, t)$ in a time complexity $O(n/t)$ that *linearly* increases with the number of inputs n and even linearly *decreases* with the number t of inputs of the gates of the module L.

Our solution strongly improves the so far known best approach of [3] that have an exponential complexity. Instead of 1731.25 s (that is almost half an hour) to solve the task for $n = 80$ and $t = 5$, our program computed this solution in less than 1 μs.

Arbitrarily we have chosen the limits $n_{\text{limit}} = 1000$ and $t_{\text{limit}} = 40$ and have computed all $p_{\min} = h'_{\min}(n, t)$ with

$$1 \leq t \leq t_{\text{limit}} = 40 \quad \text{and} \quad 1 + \frac{t \cdot (t+1)}{2} \leq n \leq n_{\text{limit}} = 1000$$

within 1 ms. Larger values of n_{limit} and fitting values of t_{limit} are possible and increase the computation time only slightly; e.g., we increased n_{limit} to $10,000$ and computed all 388,520 solutions $p_{\min}$ with $1 \leq t \leq t_{\text{limit}} = 40$ within 0.21 s.

Our approach utilizes the one-hot encoding of $S_1^n(\mathbf{x})$. Arbitrary elementary symmetric functions $S_m^n(\mathbf{x})$ with $m > 1$ can be expressed by such a one-hot encoding using AND-gates; hence, using such an additional circuit, our approach can also be used for other elementary symmetric functions $S_m^n(\mathbf{x})$ with $m > 1$.

We excluded in this chapter the analysis of modules L for $\log_2(2 \cdot t) < p < t$ (this is the region 2, introduced in Sect. 5.3) which occur for combinations of small values n and relatively large values t. In such circuits, at least one input x_i must be connected with more than two inputs of the gates of the module L. Closing this gap can be the topic of future work when results for these special cases are needed for any application.

References

1. Nagayama, S., Sasao, T., Butler, J.T.: An exact optimization method using ZDDs for linear decomposition of symmetric index generation functions. Int. Fed. Comput. Logic J. Logic Appl. **5**(9), 1849–1866 (2018)

2. Nagayama, S., Sasao, T., Butler, J.T.: A dynamic programming based method for optimum linear decomposition of index generation functions. In: Proceedings of the 49th International Symposium on Multiple-Valued Logic, pp. 144–149. IEEE, New York (2019)
3. Nagayama, S., Sasao, T., Butler, J.T.: On optimum linear decomposition of symmetric index generation functions. In: Proceedings of the 50th International Symposium on Multiple-Valued Logic. http://dx.doi.org/10.1109/ISMVL49045.2020.00-17, pp. 130–136. IEEE, New York (2020)
4. Posthoff, C., Steinbach, B.: Logic Functions and Equations—Binary Models for Computer Science, 2nd edn. ISBN 978-3-030-02419-2. Springer, Cham (2019)
5. Sasao, T.: Index generation functions: recent developments (invited paper). In: Proceedings of the 41st International Symposium on Multiple-Valued Logic, 2011, pp. 1–9. IEEE, New York (2011)
6. Sasao, T.: Linear decomposition of index generation functions. In: Proceedings of the 17th Asia and South Pacific Design Automation Conference, pp. 781–788 (2012)
7. Steinbach, B., Posthoff, C.: The last unsolved four-colored rectangle-free grid: the solution of extremely complex multiple-valued problems. In: Journal of Multiple-Valued Logic and Soft Computing, Old City Publishing, vol. 25(4–5). ISSN 1542-3980 (print), ISSN 1542-3999 (online), pp. 461–490. Philadelphia (PA), USA, (2015)

Axiomatizing Boolean Differentiation

Felix Weitkämper

1 Introduction

1.1 Our Approach

Derivative operations on Boolean algebras have been much studied since they were first described as such in the 1950s. An up-to-date textbook focused on the calculus as well as on the numerous applications of Boolean differential operations is [12], while a concise systematic treatment of the calculus can be found in Chapter 10 of [11]. However, while algebraic and numeric aspects of differentiation on Boolean algebras have been widely studied, and various fields of application have been explored, to the best of our knowledge there has been no axiomatic investigation of Boolean differentiation since [6]. We will provide an axiomatic treatment that characterizes Boolean derivatives up to isomorphism. We will discuss how this can be adapted to an axiomatization of the first-order theory and outline some potential routes for further investigation and application. We will also see how the known notions of derviative fit into the framework we propose and clarify the relationship between our axioms and those of [6].

Another major motivation of this work comes from modern model theory, where over the last decades, two areas pertinent to this research have been explored in great depth.

Firstly, there is a model theory of difference fields, which are fields equipped with an automorphism. This has been developed extensively using cutting-edge model-theoretic analysis such as the calculus of simple theories and has found deep applications in number theory and algebraic dynamics (see [1] for an introduction).

F. Weitkämper (✉)
Institut für Informatik, Ludwig-Maximilians-Universität München, München, Germany
e-mail: felix.weitkaemper@lmu.de

R. Drechsler, D. Große (eds.), *Recent Findings in Boolean Techniques*,
https://doi.org/10.1007/978-3-030-68071-8_4

We will see later that, in fact, our setting is more aligned to that of difference algebra than to the setting of differential algebra that it is often compared with.

Secondly, there has been an upsurge in research on the connection between infinite models and their finite substructures, with conference volumes such as [4] dedicated to the topic and the monograph [2] summing up a whole line of research. This is particularly relevant here since most of the application interest lies in differentiation on finite rather than infinite Boolean algebras, while the power of model theoretic methods will be felt on the infinite level.

However, while we are inspired by the work on difference fields, our setting is quite different, since firstly we are entirely concerned with characteristic 2 (since $x + x = 0$ for the symmetric difference in a Boolean algebra) and secondly the automorphisms we study are involutions rather than free automorphisms. We will see that this combination will allow us to use a very small set of axioms compared to the axiomatizations of algebraically closed fields with an automorphism in [1]. This remains true even when we move towards several derivations, while the situation for several commuting automorphisms of fields is rather complicated.

One model-theoretic advantage of the field setting over that of Boolean algebras is that algebraically closed fields are *uncountably categorical*, while Boolean algebras are *unstable*. These classifications, which we will discuss briefly in Sect. 2, mean that the most powerful tools of contemporary model theory, those from stability theory, do not apply to Boolean algebras. Therefore, we will also present an axiomatization of the reduct to a language that contains purely the symmetric difference and the derivation(s). We will show that the theory of Boolean differentiation considered in this language is in fact *totally categorical*, a very strong model-theoretic property that means that it has a unique model up to isomorphism in every infinite cardinality. This allows the direct application of methods from [2], say, to our structures.

1.2 Applications of Axiomatizing Boolean Differentiation

We will briefly outline some potential ramifications of our different levels of axiomatization. Firstly, an immediate consequence of having an axiomatization *up to isomorphism* that applies to several known notions of derivative is that those notions are indeed isomorphic. This means that any property that is preserved under isomorphism transfers immediately from one derivative to another. For instance, we will see in Theorem 5 that both vectorial and simple derivatives of switching functions in the sense of [12] fall under our axiomatization. Therefore, in order to prove something for all vectorial derivatives, it suffices to prove it for the simple derivative with respect to the first coordinate only, and it immediately generalizes to all the other notions of derivative.

From a structural point of view, a complete axiomatization lists essential properties of Boolean differentiation, since all properties of Boolean differentiation are bound to follow from the axioms. We will see that the axioms only need to pose a small number of algebraic conditions on the derivatives, which should sharpen the focus of further investigations into Boolean differentiation.

Having established an axiomatization up to isomorphism, what can be gained from an axiomatization of merely the first-order theory? One key advantage is that the first-order theory makes a connection between a single theory of derivatives on infinite Boolean algebras and the infinitely many theories of large finite Boolean Algebras with derivatives. The beginnings of this are developed in Sect. 4.

Beyond that, the connections to model theory and the work of [2] outlined above require a first-order axiomatization of the additive theory with derivatives. Exploiting this deeper connection of finite and infinite Boolean differentiation remains further work.

1.3 Outline

In the section following this introduction, we will be giving an overview of the terms and the results from model theory that we will be using in this chapter.

In the main section, we will introduce Boolean differentiation and specifically our framework for derivatives. We will provide the axiom systems for the full language and the additive reduct and prove their completeness.

In Sect. 4, we will discuss the relationship to finite algebras of logic functions equipped with derivatives, and prove elimination of quantifiers for the theories with a single derivation.

In the final section, we will discuss connections to the existing literature on Boolean differentiation. We will also highlight possible consequences of our results and point out some other putative areas for further research.

2 Model-Theoretic Fundamentals

In this section, we will rehearse the elements of classical model theory that we will need in the course of the paper. However, we will assume familiarity with the basic principles of first-order logic, such as its syntax and semantics, as well as fundamental concepts such as completeness of a theory and isomorphism of structures, which should be explained in any first textbook on logic.

Due to their traditional connection to propositional logic, Boolean algebras were among the first algebraic structures whose model theory was studied. In this work, we will refer to two classical complete theories of Boolean algebras: infinite atomic and infinite atomless Boolean algebras.

Proposition 1 *The following classes of Boolean algebras are axiomatizable by a complete first-order theory:*

1. *The theory of infinite atomless Boolean algebras*
2. *The theory of infinite atomic Boolean algebras*

We will continue with some additional definitions:

Definition 1 A first-order theory is called *categorical* in a cardinal κ if all its models of cardinality κ are isomorphic.

It is well known that the theory of infinite atomless Boolean algebras is ω-categorical, while the theory of infinite atomic Boolean algebras is not.

Categoricity is a central concept in model theory, as it implies both completeness and good model-theoretic behavior:

Proposition 2 *(Vaught's Test, Theorem 2.2.6 of [8]) If a satisfiable first-order theory with no finite models is categorical in an infinite cardinal κ, then it is complete.*

The most common measure of well-behavedness used in modern model theory is stability and the many variants of this concept, all of which have their root in Saharon Shelah's groundbreaking work on classification theory. We will refer to several steps on this scale, which we will briefly introduce here. We first need the concept of a type.

Definition 2 Let T be a complete theory, $\mathbb{M}$ a model of T, A a subset of $\mathbb{M}$, and $n \in \mathbb{N}$. Then a *(complete) n-type p of T over A* is a set of formulas with n free variables and parameters in A such that p is satisfiable, and for every such formula ϕ, either ϕ or $\neg\phi$ lies in p.

The number of types that are realized in a certain model is at the basis of one of a number of equivalent definitions of stability. However, since we will need a different formulation later, we will give that here:

Definition 3 Let T be a complete theory in a countable language.

T is called *stable* if no formula has the order property: that is, there is no model $\mathbb{M}$ of T and formula $\phi(x; y)$ such that for a sequence of pairs of tuples $(a_i; b_i)_{i<\omega}$ in $\mathbb{M}$, $\phi(a_i; b_j)$ holds if and only if $i < j$.

T is called *ω-stable* if there are only countably many types over any countable subset of a model of T.

T is called *strongly minimal* if every definable subset of any model of T is either finite or cofinite (i.e., its complement is finite).

These categories of stability are related to another in a strictly descending scale as follows:

Proposition 3 *For complete theories T in a countable language, the following strict implications hold: (i) $\Rightarrow$(ii)$\Rightarrow$(iii)$\Rightarrow$(iv), where*

(i) T is strongly minimal.
(ii) T is categorical in one (equivalently all) uncountable cardinal(s).

(iii) T is ω-stable.
(iv) T is stable.

Proof (i) implies (ii) by Proposition 6.1.12, (ii) implies (iii) by Theorem 5.2.10, (iii) implies (iv) by Proposition 6.2.11 with Theorem 6.2.14, all from [8]. □

One of the prime reasons for the usefulness of stability theory is its connection to the existence of a good dimension notion on all models of the theory. The most commonly used and strongest dimension notion is known as *Morley Rank (*alongside the associated notion of a *Morley Degree)* and is usually abbreviated as *RM.* While one can find a rigorous introduction of the notion in Chapter 6 of [8], we will here just note the relationship between the existence of a well-defined Morley Rank and the stability hierarchy given above:

Proposition 4 *Let T be a complete theory in a countable language.*

Then T is strongly minimal if and only if every model of T has Morley Rank 1 and Morley Degree 1.

If T is uncountably categorical, every model has finite Morley Rank.

T is ω-stable if and only if (every definable subset of) every model has well-defined Morley Rank.

Remark 1 T being just stable is characterized by a different, but less well-behaved rank notion being well-defined.

We will conclude our excursion to stability theory by applying the stability hierarchy to Boolean algebras.

Proposition 5 *Let T be a theory that interprets an infinite Boolean algebra. Then T is unstable.*

Proof The canonical order relation of any infinite Boolean algebra, given by $a \leq b$ iff $a = a \wedge b$, has the order property in the sense of Definition 3. □

Therefore, we will not just study the full theory of the differential Boolean calculus, but also its reduct to the additive group of the associated Boolean ring. That is an abelian group with $x + x = 0$ for all x, and thus an $\mathbb{F}_2$-vector-space. This reduct is on the opposite end of the stability spectrum:

Proposition 6 *The theory of infinite abelian groups with $x + x = 0$ for all x is strongly minimal.*

Proof Classical result of model theory; see, e.g., Section 4.5 of [5]. □

We will now continue to those concepts that help to characterize the relationship between finite and infinite structures.

First, we will introduce the concept of a generic theory, specialized to a context appropriate for our investigations:

Definition 4 Let $\mathcal{L}$ be a language and let $(\mathbb{M}_n)_{n \in \mathbb{N}}$ be a sequence of $\mathcal{L}$-structures. Let T be a complete $\mathcal{L}$-theory.

T is called the *generic theory* of $(\mathbb{M}_n)_{n\in\mathbb{N}}$ if for all $\varphi \in T$ there is an $N \in \mathbb{N}$ such that $\mathbb{M}_n \models \varphi$ for all $n > N$.

Since all finite Boolean algebras are atomic, the generic theory of the cardinality-ascending sequence of finite Boolean algebras is the theory of infinite atomic Boolean algebras.

A generic theory can be considered as a limit of the individual theories of a sequence of structures.

A different notion which may or may not coincide with a generic theory can be obtained by turning this around and considering instead the first-order theory of the limit of the structures.

The notion of limit used here is the Fraisse limit of structures, for which there are different formalizations in slightly different settings. For our purposes, we will need one that can accommodate functions as well as relations, and we find it in Section 7 of [5].

Definition 5 Let $\mathbb{M}$ be an $\mathcal{L}$-structure. Then:

$\mathbb{M}$ is called *locally finite* if any finitely generated substructure of $\mathbb{M}$ is finite.

A locally finite structure is called *uniformly locally finite* if there is a function $f : \mathbb{N} \to \mathbb{N}$ such that the substructure generated by any subset of cardinality n has cardinality at most $f(n)$.

A locally finite $\mathbb{M}$ is called *ultrahomogeneous* if every isomorphism between finite substructures extends to an isomorphism of $\mathbb{M}$.

If $\mathbb{M}$ is countably infinite, ultrahomogeneous and locally finite, it is referred to as a *Fraisse structure.*

Such a Fraisse structure is considered the *Fraisse limit* of the class of its finite substructures.

Proposition 7 *(Theorem 7.1.2 of [5]) A non-empty class of finite structures $\mathfrak{K}$ is the class of finite substructures of a Fraisse structure (i.e., has a Fraisse limit) if the following are satisfied:*

1. $\mathfrak{K}$ is closed under isomorphism.
   (a) $\mathfrak{K}$ is closed under taking substructures.
   (b) $\mathfrak{K}$ contains structures of arbitrarily large cardinalities.
   (c) Whenever A and B are in $\mathfrak{K}$, there is a C in $\mathfrak{K}$ such that both A and B can be embedded in C (Joint embedding property).
   (d) Whenever A, B_1, and B_2 are in $\mathfrak{K}$, $f_1 : A \to B_1$ and $f_2 : A \to B_2$, there are a $C \in \mathfrak{K}$ and embeddings $g_1 : B_1 \to C$ and $g_2 : B_2 \to C$ such that $g_1 \circ f_1 = g_2 \circ f_2$ (Amalgamation property).

Sometimes the generic theory of a class $\mathfrak{K}$ and the theory of the Fraisse limit coincide. For instance, the theory of infinite $\mathbb{F}_2$-vector-spaces is both the generic theory and the theory of the Fraisse limit of the class of finite $\mathbb{F}_2$-vector-spaces. For Boolean algebras, however, both notions of limit exist, but they do not coincide: While the generic theory of the class of finite Boolean algebras is the theory of infinite atomic Boolean algebras, their Fraisse limit is atomless (Classical, see Example 6.5.25 of [9]).

A very useful consequence of ultrahomogeneity is that the theory of a Fraisse structure will often be ω-categorical and admit quantifier elimination:

Proposition 8 (Theorem 7.4.1 of [5]) *Let $\mathbb{M}$ be a uniformly locally finite Fraisse structure. Then the theory of $\mathbb{M}$ is ω-categorical and admits quantifier elimination.*

As both abelian groups with $x+x=0$ for all x and Boolean algebras are uniformly locally finite, the theory of atomless Boolean algebras and the theory of infinite abelian groups with $x+x=0$ are ω-categorical and admit quantifier elimination.

3 Axiomatizing Boolean Differentiation

3.1 Boolean Functions, Rings, and Derivations

The first prerequisite for a study of structures endowed with derivative operations is to recognize the underlying algebraic nature of those structures.

We will formulate this chapter entirely in the context of *Boolean rings*, which is equivalent to that of Boolean algebras.

Definition 6 A *Boolean ring* $(\mathbb{B}, +, \cdot, 0, 1)$ is a commutative ring with unit that satisfies the following properties.

1. *Idempotency*: For any $x \in \mathbb{B}$, $x \cdot x = x$.
2. *Characteristic 2*: For any $x \in \mathbb{B}$, $x + x = 0$.

Any Boolean algebra can be made into a Boolean ring by treating $+$ as the symmetric difference (sometimes written $\oplus$ to avoid ambiguity) and $\cdot$ as the conjunction. Conversely, any Boolean ring defines a Boolean algebra, with conjunction taken as $\cdot$, disjunction as $x + y + xy$, and negation as $x + 1$. See [10] for the details.

This representation suits our purposes very well, since derivations are usually defined using the symmetric difference.

The most used derivations arise in the study of *switching functions*, that is, functions from $\{0, 1\}^n \to \{0, 1\}$ for an $n \in \mathbb{N}$. We will now formally introduce these derivations:

Definition 7 Let $n \in \mathbb{N}$, and let $f : \{0, 1\}^n \to \{0, 1\}$.

Then the *derivative of f with respect to the ith coordinate* $\delta_i(f)$ is given by the function

$$\delta_i(f) : \{0, 1\}^n \to \{0, 1\},$$

$$\delta_i(f)(a_1, \ldots, a_i, \ldots, a_n) := f(a_1, \ldots, a_i, \ldots, a_n) + f(a_1, \ldots, a_i', \ldots, a_n).$$

The *global derivative* $D(f)$ is given by $D(f)(x) = D(f)(x')$.

These derivatives have been extensively studied, and are the topic of the recent monograph [12]. In that and other work, a generalized notion of derivative that the authors call *vectorial derivative* is also introduced.

Definition 8 Let $n \in \mathbb{N}$, $f : \{0,1\}^n \to \{0,1\}$ and let $S \subseteq \{1,\dots,n\}$. Then the *vectorial derivative of f with respect to S*, $\delta_S(f)$, is given by the function

$$\delta_S(f) : \{0,1\}^n \to \{0,1\}\,,\ \delta_S(f)(a_1,\dots,a_n) := f(a_1,\dots,a_n) + f(b_1,\dots,b_n),$$

where $b_i = \begin{cases} a_i' & i \in S \\ a_i & i \notin S \end{cases}$.

In the literature, Boolean differentiation is studied mainly as an analogue to real or complex differentiation, and its algebraic properties are usually considered analogues to real or complex *differential* algebra (a remarkable exception to this being [13]).

However, while the above-mentioned derivatives are additive and factor over constants (i.e., functions whose derivative is 0), they do not satisfy the Leibniz rule of differentiation, that is, $\delta(xy) = x\delta(y) + y\delta(x) + \delta(x)\delta(y)$ rather than $\delta(xy) = x\delta(y) + y\delta(x)$, and indeed no possible notion of derivative could satisfy the classical definition of a derivation (cf. [11], Ch. 10).

In this chapter, we will instead consider Boolean differentiation as an analogue of classical *difference* algebra, which studies automorphisms of the real or complex field. This possibility arises from the following observation:

Proposition 9 *In all of the cases above, the map $f \to \sigma(f)$, $\sigma(f)(a_1,\dots,a_n) := f(b_1,\dots,b_n)$ is an involution of the Boolean ring of functions from $\{0,1\}^n \to \{0,1\}$.*

Proof We need to show that σ respects addition and multiplication and that $\sigma^2 = \sigma$.

1. $\sigma(f+g)(a_1,\dots,a_n) = (f+g)(b_1,\dots,b_n) = f(b_1,\dots,b_n)+g(b_1,\dots,b_n) = \sigma(f)(a_1,\dots,a_n) + \sigma(b)(a_1,\dots,a_n)$.
2. Similarly for multiplication.
3. $\sigma^2(f)(a_1,\dots,a_n) = f(b_1,\dots,b_n)$, where $b_i = \begin{cases} a_i'' & i \in S \\ a_i & i \notin S \end{cases}$.

   But as $a_i'' = a_i$, $b_i = a_i$ for all i.

□

As $\delta(f) = f + \sigma(f)$, $\sigma(f) = f + \delta(f)$ and we can (and will) therefore study derivations and their associated involutions interchangeably.

3.2 A Complete Axiomatization

In the light of Proposition 9, we can choose between using a derivation δ or an involution σ in our language, and whether to include the operations of a Boolean algebra or the ring operations. For the sake of consistency with the notion of Boolean *differentiation*, we will officially present our axiomatization in the following languages:

Definition 9 For $n \in \mathbb{N}$, let $\mathcal{L}_n$ be the language consisting of the binary operations $+$ and $\cdot$, the constant symbols 0 and 1 and the unary functions $\delta_1, \ldots, \delta_n$.

Let $\mathcal{L}_n^+$ be the reduct of this language, where the conjunction $\cdot$ and the constant 1 are omitted.

Whenever R is an $\mathcal{L}_n$ or $\mathcal{L}_n^+$ structure, let $\sigma_n := \delta_n + \text{id}$.

In $\mathcal{L}_1$ and $\mathcal{L}_1^+$, we usually write δ and σ for δ_1 and σ_1.

For clarity of exposition, we will begin by providing a complete axiomatization of the Boolean derivative on $\mathcal{L}_n^+$ and then extending it to a complete axiomatization on $\mathcal{L}_n$.

Definition 10 Let T_1^+ be the following $\mathcal{L}_1^+$ theory:

1. V is an abelian group of characteristic 2, that is, an abelian group with the property that $\forall x(x + x = 0)$.
2. σ is an involution of groups.
3. δ is complete, that is, $\forall y(\delta(y) = 0 \Rightarrow \exists x(\delta(x) = y))$.

We will not only show that T_1^+ is complete when restricted to infinite models, but moreover, we will show that it is categorical in every infinite cardinal:

Theorem 1 *T_1^+ is categorical in all infinite cardinals. Its infinite models form a complete ω-stable elementary class.*

The proof of Theorem 1 will go through two Lemmas. First, though, a simple observation that we will use throughout and which justifies the formulation of the completeness axiom:

Remark 2 Let V be an abelian group of characteristic 2 and σ an involution of groups. Then $\forall x \in V : \delta(\delta(x)) = 0$.

Proof $\delta(\delta(x)) = \delta(x+\sigma(x)) = (x+\sigma(x))+\sigma(x+\sigma(x)) = x+\sigma(x)+\sigma(x)+x = 0$. □

Lemma 1 *Let V, V' be free finite-dimensional k-modules over a ring k and let $F : V \to V'$ be a linear isomorphism. Let $f : V \to V$ and $f' : V' \to V'$ be linear endomorphisms. Let **a** be a basis for V and M a matrix representing f with respect to **a**. Then F is an isomorphism of the structures enriched by a function symbol for f on V and f' on V' iff M is the matrix representation of f' with respect to $\overrightarrow{F(a)}$.*

Proof It suffices to show that $\forall x(F(f(x)) = f'(F(x)))$. So let $x \in V$ and let $\mathbf{v}$ be the k-vector representing x with respect to $\mathbf{a}$. Then $\mathbf{v}$ is also the k-vector representing $F(x)$ with respect to $\overrightarrow{F(a)}$. Thus

$$F(f(x)) = F(M\mathbf{va}) = M\mathbf{v}\overrightarrow{F(a)} = f'(\mathbf{v}\overrightarrow{F(a)}) = f'(F(\mathbf{va})) = f'(F(x)).$$

□

In order to apply Lemma 1 to our structures, we will prove another lemma.

Lemma 2 *Let $(V, +, 0, \delta)$ be a model of T_1^+. Then $(V, +, 0)$ is an $\mathbb{F}_2$ vector space and the following holds:*

1. $V = \bigoplus_{i=1}^{\kappa} U_i$ *for a cardinal κ, where each U_i is a 2-dimensional δ-invariant subspace on which δ can be represented by the matrix* $\begin{pmatrix} 0 & 0 \\ 1 & 0 \end{pmatrix}$.
2. *V has cardinality 2^{2n} for an $n \in \mathbb{N}$ or infinite cardinality.*

Proof The proof will proceed in steps.

First, as V is an abelian group, being of characteristic 2 is equivalent to being an $\mathbb{F}_2$ vector space.

Let K be the kernel of the group and thus $\mathbb{F}_2$-vector-space-homomorphism δ. Let $(b_i | i \in I)$ be an $\mathbb{F}_2$-basis for K and let $(a_i | i \in I)$ be such that $\delta(a_i) = b_i$. We claim that $V = \bigoplus_{i \in I} \langle a_i, b_i \rangle$ is a decomposition as required in the statement of the lemma. So, we have to show (a) that $V = \sum_{i \in I} \langle a_i, b_i \rangle$, (b) that the sum is direct, and (c) that each $\langle a_i, b_i \rangle$ satisfies the requirements of the lemma.

(a): Let $x \in V$. Then by Remark 2 $\delta(x) \in K$ and thus $\delta(x) = \sum_{j \in J} b_j$. Observe that $\delta(x + \sum_{j \in J} a_j) = \sum_{j \in J} b_j + \sum_{j \in J} b_j = 0$ and thus that $x + \sum_{j \in J} a_j \in K$. But as by definition $K \subseteq \sum_{i \in I} \langle a_i, b_i \rangle$ and $\sum_{j \in J} a_j \in \sum_{i \in I} \langle a_i, b_i \rangle$, we also obtain $x \in \sum_{i \in I} \langle a_i, b_i \rangle$.

(b): We need to show that $\sum_{j \in J} u_j = 0 \Rightarrow u_j = 0$ for all $j \in J$. But by definition $\sum_{j \in J} u_j = \sum_{k \in K} a_k + \sum_{l \in L} b_l$. We see that $\delta(\sum_{k \in K} a_k + \sum_{l \in L} b_l) = \sum_{k \in K} b_k$ and since $(b_i | i \in I)$ is a basis for K, this implies that $K = \emptyset$. Then $\sum_{l \in L} b_l = 0$, which however implies that $L = \emptyset$ by the same argument.

(c): We have already seen that each $\langle a_i, b_i \rangle$ is 2-dimensional, so it remains to show that $\delta(a_i) = b_i$ and that $\delta(b_i) = 0$. But that is just the definition of the a_i and b_i.

This shows the first clause of the Lemma; the second clause follows from the first clause together with additivity of dimension in free sums and the fact that $|V| = 2^{\dim_{\mathbb{F}_2}(V)}$. □

Remark 3 In fact, one can extend any linearly independent system $\mathbf{w_i}$ in the kernel together with any $\mathbf{v_i}$ with $\delta(v_i) = w_i$ into a representation with respect to which the lemma holds.

We can now proceed to prove Theorem 1.

Proof Let $(V, +, \delta)$ and $(V', +, \delta')$ be two models of T_1^+ of cardinality $\kappa \geq \omega$. Then by Lemma 2, $V = \bigoplus_{i=1}^{\kappa} U_i$ and $V' = \bigoplus_{i=1}^{\kappa} U_i'$ with the properties mentioned there. We define a linear bijection $F : V \to V'$ by defining linear bijections $F_i : U_i \to U_i'$ for each i. Let (a_i, b_i) and (a_i', b_i') be bases for U_i and U_i', respectively, for which δ has the matrix representation $\begin{pmatrix} 0 & 0 \\ 1 & 0 \end{pmatrix}$. Then let $F_i(a_i) = a_i'$ and $F_i(b_i) = b_i'$. Clearly, F_i defines an isomorphism of vector spaces, and by Lemma 1, $F_i(\delta(x)) = \delta'(F(x))$. We will now define $F(x) = F(\sum u_j) := \sum F_j(u_j)$. This is clearly a well-defined linear bijection. It thus only remains to show that $F(\delta(x)) = \delta'(F(x))$:

$$F(\delta(x)) = \sum F_j(\delta(u_j)) = \sum \delta'(F_j(u_j)) = \delta'(\sum F_j(u_j)) = \delta'(F(x)).$$

Therefore, T_1^+ is categorical in all infinite cardinals. By the discussion in Sect. 2, this implies that the first-order theory of the infinite models of T_1^+ is both complete and ω-stable (since it is uncountably categorical).

This categoricity result unlocks powerful model-theoretic tools for Boolean differential groups, which we will briefly discuss in the final section. Here we will now adapt our axiomatization to give a complete first-order theory of Boolean differentiation which takes full account of the ring structure. □

Definition 11 Let K be a Boolean ring, and T_K a complete first-order theory of Boolean algebras expressed in the language of Boolean rings. Then T_1^K is the following theory in the language $\mathcal{L}_1$:

1. σ is an involution of Boolean rings.
2. $\ker(\delta) \models T_K$.
3. δ is complete, i.e., there is a $z \in V$ such that $\delta(z) = 1$.

Remark 4 We remark that we found it rather surprising that one could obtain a complete axiomatization by just adding a finite number of axioms to the ones regarding K. This seems to be entirely due to the fact that one can define the ring structure on V from the ring structure on the constants (see below).

We will adopt a different and possibly more straightforward strategy to proving completeness of the first-order theory here, extending isomorphisms between kernels to isomorphisms between the models of T_1^K. First, we give a more concrete characterization of δ being complete:

Proposition 10 *Let V be a model of T_1^K for a Boolean ring K. Then V is a free* $\ker(\delta)$*-algebra on two generators* $(1, z)$ *and δ is a* $\ker(\delta)$*-algebra-morphism given by* $\delta(z) = 1$ *and* $\delta(1) = 0$.

Proof Let z be as in the definition of T_1^K.

(a) $(1, z)$ generate V. Indeed, let $x \in V$ be arbitrary. Then $\delta(x) \in \ker(\delta)$ and $\delta(x) = \delta(\delta(x)z)$ by $\ker(\delta)$-linearity. Thus, $x + \delta(x)z \in \ker(\delta)$ and therefore $x = (x + \delta(x)z) + \delta(x)z$ is the required representation.
(b) $(1, z)$ generate V freely. Indeed, if $a + bz = 0$ for some $a, b \in K$, then $\delta(a + bz) = b = 0$ and thus also $a = 0$.

□

Now we can prove the extension of isomorphisms.

Proposition 11 *There is a one-to-one correspondence between isomorphism classes of Boolean algebras K and isomorphism classes of models of T_1^K.*

Proof Let K be a Boolean algebra and V a free K-algebra on 2 generators. Then by Lemma 1, the condition $\delta(z) = 1$ and $\delta(1) = 0$ uniquely determines V as a K-algebra up to isomorphism.

So let $f : V \to V'$ be an isomorphism of K-algebras respecting δ. We claim that f is in fact an isomorphism of Boolean rings. So let $(k_1 + k_2 z)$ and $(k_1' + k_2' z)$ be elements of V. Then

$$
\begin{aligned}
& f\left((k_1 + k_2 z) \cdot \left(k_1' + k_2' z\right)\right) = f\left(k_1 k_1' + \left(k_2 k_1' + k_1 k_2' + k_2 k_2'\right) z\right) \\
&= f(k_1) f\left(k_1'\right) + \left(f(k_2) f\left(k_1'\right) + f(k_1) f\left(k_2'\right) + f(k_2) f\left(k_2'\right)\right) f(z) \\
&= f(k_1 + k_2 z) f\left(k_1' + k_2' z\right)
\end{aligned}
$$

Therefore f is actually an isomorphism of Boolean rings as required. □

We can reformulate this as a complete axiomatization result in its own right:

Corollary 1 *For any Boolean algebra K, the following three axioms characterize a Boolean derivative with kernel K up to isomorphism:*

1. *σ is an involution of Boolean rings.*
2. $\ker(\delta) \cong K$.
3. *δ is complete, i.e., there is a $z \in V$ such that $\delta(z) = 1$.*

It follows from the above that whenever the theory of K is ω-categorical, then so is T_1^K. In particular, when K is an infinite atomless Boolean algebra, then T_1^K is ω-categorical and therefore complete. In fact, T_1^K is complete regardless of K, and this can be seen using any of a number of classical model-theoretic techniques.

Theorem 2 *Let T_K be any complete theory of Boolean rings. Then the theory T_1^K is complete.*

Proof We sketch a proof using ultraproducts (see Section 9.5 of [5] for an introduction), since that most easily generalizes to several derivations. Let A and B be models of T_1^K, and let K_A and K_B be their respective kernels. Then $K_A \equiv K_B$ and we want to show that $A \equiv B$ also. By the Keisler-Shelah Theorem, K_A and K_B have isomorphic ultrapowers $U(K_A) \simeq U(K_B)$. Using the same index set and the same ultrafilter, we can take the ultrapowers of $U(A)$ of A and $U(B)$ of B. Then the kernel of $U(A)$ is isomorphic to $U(K_A)$ and the kernel of $U(B)$ is isomorphic to $U(K_B)$. Thus, the kernels are isomorphic to each other and by Proposition 11 $U(A)$ and $U(B)$ are also. Therefore, A and B must have been elementarily equivalent. □

We will now extend the characterizations above to several derivatives.

In the following, we will use the shorthand $\delta_J^{|J|}$ to mean the $|J|$-fold derivative with respect to all δ_j, $j \in J$; for instance, $\delta^n_{\{1,\dots,n\}}(x) = \delta_1\delta_2\dots\delta_n(x)$ and $\delta^1_{\{j\}} = \delta_j$. (We add the cardinality superscript to avoid confusion with the vectorial derivative from Definition 8.)

Definition 12 Let T_n^+ be the following $\mathcal{L}_n^+$ theory:

1. V is an abelian group of characteristic 2, that is, an abelian group with the property that $\forall x(x + x = 0)$.
2. $\sigma_1, \dots, \sigma_n$ are commuting involutions of groups.
3. $\{\delta_1, \dots, \delta_n\}$ is complete, that is,

$$\forall y(\delta_1(y) = 0 \wedge \delta_2(y) = 0 \wedge \dots \wedge \delta_n(y) = 0 \Rightarrow \exists x(\delta_1\delta_2\dots\delta_n(x) = y)).$$

We will now provide an analogue to Lemma 2 to prove the categoricity of T_n^+ in each uncountable cardinal.

Lemma 3 *Let $(V, +, 0, \delta_1, \dots, \delta_n)$ be a model of T_n^+. Then $(V, +, 0)$ is an $\mathbb{F}_2$ vector space and the following holds:*

1. *$V = \bigoplus_{i=1}^{\kappa} U_i$ for a cardinal κ, where each U_i is a 2^n-dimensional δ-invariant subspace which has a basis $(a_{i,J} | J \subseteq \{1, \dots, n\})$ such that the following holds: $\{\langle\{a_{i,J}, a_{J\cup\{j\}}\}\rangle \,|\, j \notin J\}$ is a decomposition of U_i in the sense of Lemma 2 with respect to δ_j.*
2. *V has cardinality $2^{2^n m}$ for an $m \in \mathbb{N}$ or infinite cardinality.*

Proof The proof will proceed in steps.

First, as V is an abelian group, being of characteristic 2 is equivalent to being an $\mathbb{F}_2$ vector space.

Let (b_i) be a basis for $\bigcap_{j=1}^{n} K_i$, where $K_j := \ker(\delta_j)$. Then choose (a_i) such that $\delta_1\delta_2\dots\delta_n(a_i) = b_i$. Let $a_{i,J} := \delta_J^{|J|}(a_i)$. We claim that this satisfies the requirements, and we will prove this by induction. The case $n = 1$ has been shown in Lemma 2. So assume true for n. It is easy to see that K_{n+1} is a model of T_n^+. Therefore, by the induction hypothesis, $(a_{i,J} | J \subseteq \{1, \dots, n + 1\}, n + 1 \in J)$ is a basis for K_{n+1} as required. But then by Lemma 2, $(a_{i,J}|J \subseteq \{1, \ldots, n\})$ is a basis of V with exactly the properties described in clause 1.

This shows the first clause of the Lemma; the second clause follows from the first clause together with additivity of dimension in free sums and the fact that $|V| = 2^{\dim_{\mathbb{F}_2}(V)}$. □

We can now deduce the completeness and indeed the total categoricity of T_n^+ just as we did for T_1^+:

Theorem 3 *T_n^+ is categorical in all infinite cardinals. Its infinite models form a complete ω-stable elementary class.*

Proof Just as in the proof of Theorem 1, the linear bijection induced by the bases given by Lemma 3 is an $\mathcal{L}_n$-isomorphism by Lemma 1. □

We will now finally provide an axiomatization of the complete theory of several derivations on Boolean rings:

Definition 13 Let K be a Boolean ring, and T_K a complete first-order theory of Boolean algebras expressed in the language of Boolean rings. Then T_n^K is the following theory in the language $\mathcal{L}_n$:

1. $\sigma_1, \ldots, \sigma_n$ are commuting involutions of Boolean rings.
2. $\bigcap_{i=1}^{n} \ker(\delta_i) \models T_K$.
3. $\{\delta_1, \ldots, \delta_n\}$ is complete, that is,

$$\exists x(\delta_1\delta_2 \ldots \delta_n(x) = 1)).$$

The proof will again be preceded by a proposition giving a more concrete representation.

Proposition 12 *Let V be a model of T_n^K for a Boolean ring K. Then V is a free $\bigcap_{i=1}^{n} \ker(\delta_i)$-algebra on 2^n generators given by $\{a_J := \delta_J^{|J|}(a)|J \subseteq \{1, \ldots, n\}\}$ for any $a \in V$ with $\delta_1\delta_2 \ldots \delta_n(a) = 1$.*

Proof By induction on n. The case $n = 1$ is part of Proposition 10. So assume it true for n and choose any model V of T_{n+1}^K and any $a \in V$ with $\delta_1\delta_2 \ldots \delta_{n+1}(a) = 1$.

We will now show that it is a generating system for V. So let $x \in V$. We will proceed by induction on the smallest number m such that the m-fold derivative $\delta_{\{1,\ldots,m\}}^m(x) \in \bigcap_{i=1}^{n} \ker(\delta_i)$. If $m = 0$ then $x \in \bigcap_{i=1}^{n} \ker(\delta_i)$ itself. So assume true for m. Then if $\delta_{\{1,\ldots,m+1\}}^m x \in \bigcap_{i=1}^{n} \ker(\delta_i)$, $x = (\delta_{\{1,\ldots,m+1\}}^m x) \cdot \delta_{\{1,\ldots,n\}\setminus\{1,\ldots,m+1\}}^{n-(m+1)} a + y$, $y := ((\delta_{\{1,\ldots,m+1\}}^m x)\delta_{\{1,\ldots,n\}\setminus\{1,\ldots,m+1\}}^{n-(m+1)} a + x)$. Here $\delta_{\{1,\ldots,m+1\}}^m y = 0$ and thus $\delta_{\{1,\ldots,m\}}^m y \in \bigcap_{i=1}^{m+1} \ker(\delta_i)$. □

Proposition 13 *Let K be a Boolean ring. Then there is exactly one model of T_K^n up to isomorphism with $\bigcap_{i=1}^{n} \ker(\delta_i) = K$.*

Proof We will prove the theorem by induction on n. The case $n = 1$ is exactly Proposition 11. So assume it true for T_K^n. We will now show it for T_K^{n+1}. Let a be the witness of clause 3 of the definition and let $a_J := \delta_J^{|J|}(a)$. Then we claim that the isomorphism of K-modules induced by a_J is an $\mathcal{L}_{n+1}$-isomorphism. By the induction hypothesis, it is an isomorphism of the obvious $\mathcal{L}_n$-structures on $K_i := \ker(\delta_i)$ for each derivation δ_i. However, since $\delta_i(a_{\{1,\ldots,i-1,i+1,\ldots,n\}}) = 1$, another application of Proposition 11 shows that we actually have an isomorphism of Boolean rings which also respects δ_i. Since i was arbitrarily chosen, this finishes the proof. □

Just as for T_1^K, we can reformulate this as an explicit axiomatization result and deduce completeness of the first-order theory:

Corollary 2 *For any Boolean algebra K, the following three axioms characterize Boolean derivatives with kernel K up to isomorphism:*

1. $\sigma_1, \ldots, \sigma_n$ *are commuting involutions of Boolean rings.*
2. $\bigcap_{i=1}^{n} \ker(\delta_i) \cong K$.
3. $\{\delta_1, \ldots, \delta_n\}$ *is complete, that is,*

$$\exists x(\delta_1\delta_2 \ldots \delta_n(x) = 1)).$$

Theorem 4 *Let T_K be any complete theory of Boolean rings, and let $n \in \mathbb{N}$. Then the theory T_n^K is complete.*

4 Relationship to Finite Models and Immediate Consequences

In this section, we will be connecting the complete theories from Sect. 3.2 with the examples of Boolean differentiation studied in the literature.

In particular, we will show that the theories we have introduced can be naturally characterized as the generic or as the limit theories of groups or rings of switching functions equipped with the derivatives introduced in Sect. 3.1.

To facilitate notation, we introduce

Definition 14 Let $\mathbb{S}_n$ be the *Boolean ring of switching functions in n variables,* that is, the Boolean ring made up of all mappings $f : \{0, 1\}^n \to \{0, 1\}$, equipped with the ring structure from Sect. 3.1. Let $\mathbb{S}_n^+$ be the additive group reduct of $\mathbb{S}_n$.

Theorem 5 *The theory of infinite models of T_1^+ is the generic theory of the class $\{\mathbb{S}_n^+ | n \in \mathbb{N}\}$, where each switching algebra is equipped with any of the derivatives of Sect. 3.1.*

Proof By the results at the end of Sect. 3.1, each of the structures mentioned is a model of T_1^+. Clearly, $|\mathbb{S}_n| \geq n$ and thus the additional infinity axioms are generically true in the class too. So, the theory of infinite models of T_1^+ is a subset of the generic theory. However, as the theory is complete by Theorem 1, it is the generic theory. □

The equivalent result for the theory of Boolean rings is obtained in a very similar way; however, one has to choose a Boolean algebra that models the generic theory of finite Boolean algebras.

Theorem 6 *The theory T_1^K, where K is an infinite atomic Boolean algebra, is the generic theory of the class $\{\mathbb{S}_n | n \in \mathbb{N}\}$, where each switching algebra is equipped with any of the derivatives of Sect. 3.1.*

Proof By the discussion following Definition 4, the theory of K is the generic theory of the class $\{\mathbb{S}_n | n \in \mathbb{N}\}$ as Boolean rings. Since $|\ker(\delta)| = \sqrt{|\mathbb{S}_n|}$ for all derivations mentioned in Sect. 3.1, the theory of $\ker(\delta)$ will indeed be generically T_K. The remainder of the axioms are clear. As T_1^K is complete by Theorem 11, we can conclude that T_1^K is the generic theory. □

The theorems above show that we have indeed given a characterization of the asymptotic theory of switching functions—so although our results and methods have focused on infinite models, they can be used to study the derivations on arbitrarily large finite switching algebras that have spawned such a large literature.

They also generalize to the theories with several derivations, when one considers derivations that are *linearly independent* in the sense of [12], but we will omit the generalization of the proofs here for brevity. One example of such linearly independent derivations are the single derivations $\delta_1, \ldots, \delta_n$ on $\mathbb{S}_n$. We have

Theorem 7 *The theory of infinite models of T_n^+ is the generic theory of the class $\{\mathbb{S}_i^+ | i \geq n\}$, where each switching algebra is equipped with the single derivatives δ_1 to δ_n.*

The theory T_n^K, where K is an infinite atomic Boolean algebra, is the generic theory of the class $\{\mathbb{S}_i | i \geq n\}$, where each switching algebra is equipped with the single derivatives δ_1 to δ_n.

We will now move on to characterize the theories we have constructed as the complete theories of limit structures. This gives us more information about their model theory and provides a concrete structure into which the finite switching algebras can be uniquely embedded up to isomorphism. We can take the limit over the same classes we have considered above. In the additive case, we will obtain exactly the same theory, as the underlying theory of infinite $\mathbb{F}_2$-vector spaces is both generic and limit theory of the finite $\mathbb{F}_2$-vector spaces. In the full Boolean ring case, however, we will have to change the Boolean ring K under consideration since the limit structure of finite Boolean rings is the countable atomless Boolean algebra and not a countable atomic Boolean algebra.

Theorem 8 *Let $\mathfrak{C}$ be the class of all substructures of a member of the class $\{\mathbb{S}_i^+ | i \in \mathbb{N}\}$, where each switching algebra is equipped with any of the derivatives of Sect. 3.1. Then $\mathfrak{C}$ is a Fraisse class and its limit structure is the unique countably infinite model of T_1^+.*

Proof We will go through the requirements of a Fraisse class one by one.

1. Closure under isomorphisms is clear.
2. Closure under substructure is guaranteed by our definition as being substructures of a certain other class of structures.
3. It contains arbitrarily large structures, as S_n lies in $\mathfrak{C}$.
4. We can always consider the larger of the two indices of the structures that they embed into.
5. Consider the situation of the amalgamation condition. As we can embed B_1 and B_2 into S_i and S_j respectively, we can assume without loss of generality that B_1 and B_2 are in $\{\mathbb{S}_i^+ | i \in \mathbb{N}\}$. Let δ denote the derivations. Without loss of generality, let the index of B_1 be at most the index of B_2. We will first build a basis for A, which we will then extend to bases for B_1 and B_2 in such a way that a natural embedding between the bases defines an embedding from B_1 into B_2. Start with a basis for $\delta(A) \subseteq A$. This can be extended to a basis for $\ker_A(\delta)$, and that in turn to bases $(\mathbf{k})$ and $(\mathbf{k}')$ of B_1 and B_2, respectively. By (the proof of) Lemma 2, $\mathbf{k}$ and $\mathbf{k}'$ together with any choice of preimages of $\mathbf{k}$ and $\mathbf{k}'$ define bases for B_1 and B_2. We can therefore choose the preimages in such a way that the preimage will be chosen from A wherever A contains such a preimage. We argue that the bases $\mathbf{b}$ and $\mathbf{b}'$ obtained in that way contain a basis for A. Indeed, by the standard kernel-image decomposition in linear algebra, the dimension of A is equal to the dimension of the image plus the dimension of the kernel, and the number of preimage elements that could be chosen from A is exactly the dimension of the image. So consider the embedding from B_1 into B_2 that is induced by mapping $\mathbf{b}$ to $\mathbf{b}'$ in an appropriate way. Then by Lemma 1, this is an isomorphism onto its image, i.e., an embedding, and it respects A as required by clause 5.

□

Thus, we could also define the theory of infinite models of T_1^+ as the theory of the Fraisse limit of all finite switching algebras equipped with a derivation.

The analysis also yields quantifier elimination as a consequence:

Corollary 3 *The theory of infinite models of T_+^1 has quantifier elimination.*

Proof The substructure generated by a subset A of a differential group is the group generated by $A \cup \sigma(A)$. Thus, since the group reduct is uniformly locally finite, so is the Boolean differential group.

Thus, the result follows from Theorem 8 by Proposition 8. □

Considering the theory with the full Boolean algebra structure, we obtain a representation for T_1^K, where K is the countable atomless Boolean algebra.

Theorem 9 *Let $\mathfrak{C}$ be the class of all substructures of a member of the class $\{\mathbb{S}_i | i \in \mathbb{N}\}$, where each switching algebra is equipped with any of the derivatives of Sect. 3.1. Then $\mathfrak{C}$ is a Fraisse class and its limit structure is the unique countably infinite model of T_1^K, where K is the countable atomless Boolean algebra.*

Proof As in the proof of Theorem 8, Clauses 1–4 are easily verified. We therefore consider the situation of the amalgamation property, and again we can assume without loss of generality that B_1 and B_2 are in $\{\mathbb{S}_i | i \in \mathbb{N}\}$ and that the index of B_1 is at most the index of B_2. Due to the corresponding property for pure Boolean algebras, we can furthermore assume that $\ker(\delta)_{B_1} \subseteq \ker(\delta)_{B_2}$ and that $(\ker(\delta) \cap A)_{B_1} = (\ker(\delta) \cap A)_{B_2}$. By the analysis in Chapter 3 of [12], $\delta(A)$ is itself a lattice of functions. In particular, $\delta(A)$ has a maximum, say $\alpha \in A$. Let $x \in A$ be chosen with $\delta(x) = \alpha$. Choose z_1 and z_2 in B_1 and B_2, respectively, such that $\delta(z_1) = 1$ and $\delta(z_2) = 1$. We will define an embedding $\iota : B_1 \to B_2$ by setting ι to be the identity on $\ker(\delta)$ and choosing a value of $\iota(z_1)$. If $\delta(\iota(z_1)) = 1$, then ι is an embedding of Boolean differential algebras. So consider $x = \alpha z_1 + a$ in B_1 and $x = \alpha z_2 + b$ in B_2, where $\delta(a) = \delta(b) = 0$. Then we set $\iota(z_1) := z_2 + a + b$. This defines an embedding of Boolean differential algebras since $\delta(z_2 + a + b) = 1 + 0 + 0 = 1$. We thus have to show that for all $y \in A$, $\iota(y_{B_1}) = \iota(y_{B_2})$. First, we will see that this holds for x, and we will derive an auxiliary result:

$$\begin{aligned} x\alpha &= \alpha z_1 + a\alpha = x + (\alpha + 1)a \\ &= \alpha z_2 + b\alpha = x + (\alpha + 1)b \\ &\Rightarrow (\alpha + 1)a = (\alpha + 1)b \\ &\Rightarrow (\alpha + 1)(a + b) = 0 \\ &\Rightarrow \alpha(a + b) = a + b \end{aligned}$$

So $\iota(x) = \alpha(z_2 + a + b) + a = \alpha z_2 + b$ as required. So now consider $y \in A$ arbitrary. Then $y = \beta z_1 + c = \beta\alpha z_1 + c = \beta(\alpha z_1 + a) + \beta a + c = \beta x + \beta a + c$. Since ι is the identity on the kernel elements β, a, and c and $\iota(x_{B_1}) = \iota(x_{B_2})$ it follows that $\iota(y_{B_1}) = \iota(y_{B_2})$ as required.

Therefore the theory has a Fraisse limit.

Since ultrahomogeneity of the whole structure also implies ultrahomogeneity of the kernel, the kernel must be the countable atomless Boolean algebra. □

Just as for the additive theory, we can now conclude a quantifier elimination result:

Corollary 4 *Let K be an atomless Boolean algebra. Then the theory T_1^K admits quantifier elimination.*

Proof Just as Corollary 3 follows from Theorem 8. □

We will conclude this section by outlining how such results might be combined to investigate Boolean differentiation in large switching algebras. We will focus on the simplest case, T_+^1, for which Fraisse theory and generic theory coincide.

The approach rests on three results: By Theorem 5, the theory of infinite models of T_+^1 is the generic theory of finite switching algebras. By Corollary 3, this theory also eliminates quantifiers. Finally, since T_+^1 is finitely axiomatizable and the theory of its infinite models is complete, it is also decidable. We can put these results together and arrive at the following two-step procedure to decide whether a given tuple of switching functions in a large switching algebra satisfies any $\mathcal{L}_1^+$-formula $\varphi(\mathbf{x})$:

1. Find a a quantifier-free formula $\varphi_0(\mathbf{x})$ such that $\varphi_0(\mathbf{x})$ is equivalent to $\varphi(\mathbf{x})$ on all infinite models of T_+^1. Since their theory is decidable, $\varphi_0(\mathbf{x})$ can be determined effectively.
2. Check whether the given tuple of switching functions $\mathbf{f}$ satisfies $\varphi_0(\mathbf{x})$. Since the theory of infinite models of T_+^1 is generic, $\varphi_0(\mathbf{f})$ iff $\varphi(\mathbf{f})$ whenever $\mathbf{f}$ is from a sufficiently large switching algebra.

Since quantifier-free formulas do not reference any other objects of the algebra, it can be checked without regard to the switching algebra from which $\mathbf{f}$ is taken but merely by inspecting $\mathbf{f}$ itself.

5 Future Applications and Perspectives

In this section, we will briefly discuss the connection between the first-order theory as presented here and Kühnrich's abstract notion of a Boolean derivative (see Chapter 10 of [11]). We will then explore potential applications and directions for further research.

While this is to the best of our knowledge the first analysis of the first-order theory of Boolean differentiation or of axioms complete up to isomorphism, there has certainly been some work on a more general framework for the different notions of derivative suggested in the literature. One such framework, which has been proposed by Martin Kühnrich [6], is presented in the chapter on Boolean differentiation in [11]:

Definition 15 Let B be a Boolean ring and let $d : B \to B$. Then d is called a *(Kühnrich) differential operator* if the following hold:

1. For all $x \in B$, $d(d(x)) = 0$.
2. For all $x \in B$, $d(x + 1) = dx$.

3. For all $x, y \in B$, $d(xy) = xd(y) + yd(x) + d(x)d(y)$.

Since Kühnrich's axioms do not include any notion of completeness, they are essentially weaker than the theory presented here. In fact, Kühnrich's differential operator has a simple characterization in terms of involutions:

Proposition 14 *Let B be a Boolean ring and let $d : B \to B$. Then d is a (Kühnrich) differential operator if and only if $\sigma : B \to B$, $\sigma(x) = x + d(x)$, is an involution of Boolean rings.*

Proof "⇒": We will verify that σ respects addition, multiplication, 0 and 1.

1. d respects addition by Proposition 10.2.1 of [11]. Thus $\sigma(x + y) = x + y + d(x + y) = x + d(x) + y + d(y) = \sigma(x) + \sigma(y)$.
2. $\sigma(xy) = xy + (xd(y) + yd(x) + d(x)d(y)) = (x + d(x))(y + d(y)) = \sigma(x)\sigma(y)$.
3. $d(0) = d(1) = 0$ by Proposition 10.2.1 of [11]. Thus $\sigma(0) = 0 + 0 = 1$ and $\sigma(1) = 0 + 1 = 1$.

"⇐": Let σ be an involution and $d(x) := x + \sigma(x)$. We will verify Kühnrich's axioms for d.

1.

$$d(d(x)) = d(x) + \sigma(d(x)) = x + \sigma(x) + \sigma(x) + \sigma(\sigma(x)) = x + \sigma(x) + \sigma(x) + x = 0.$$

2.

$$d(x + 1) = x + 1 + \sigma(x) + \sigma(1) = x + \sigma(x) = d(x).$$

3.

$$\begin{aligned} & xd(y) + yd(x) + d(x)d(y) \\ &= x(y + \sigma(y)) + y(x + \sigma(x)) + (x + \sigma(x))(y + \sigma(y)) \\ &= xy + x\sigma(y) + xy + y\sigma(x) + xy + x\sigma(y) + y\sigma(x) + \sigma(x)\sigma(y) \\ &= xy + \sigma(x)\sigma(y) = d(xy). \end{aligned}$$

□

This holds completely analogously for the "Boolean differential algebras of order k" that are introduced in Definition 10.2.2 of [11]; they are exactly characterized by inducing k commuting involutions of B.

This characterization suggests the question of the exact relationship between Kühnrich's operators and the models of the theories introduced here. In particular, it is clear that every substructure of a model of T_1^K (T_n^K) is a differential operator (algebra) in this sense. But does the converse hold? By Corollary 6.5.3 of [5], this is equivalent to the question of whether Kühnrich's axioms axiomatize the universal theory of $T_1^K (T_n^K)$.

A particular interest lies in the connections between the finite structures that are studied in the literature and the complete theories of infinite structures expounded here. For the particular case of the additive reduct, this is especially alluring, since their infinite models form a totally categorical theory. The connection between totally categorical theories and their finite substructures is the subject of a deep model-theoretic analysis around so-called *smoothly approximable structures* as discussed for instance in [7] and [2]. In particular, the Morley rank of a definable set in the theory of infinite models determines the approximate size of the respective definable subset of a finite model in a precise and uniform manner.

Of course, stability theory brings a host of interrelated concepts in its own right too, and investigating these notions with respect to additive Boolean differentiation would be an important contribution towards bringing the theory of difference algebra in the Boolean case to a similar level as the more widely studied difference algebra over fields.

Furthermore, it would be very interesting to extend T_n^K and T_n^+ to countably infinitely many derivations. Then, one would have one single theory encompassing switching functions of arbitrary sizes and their derivatives. Using more sophisticated model-theoretic techniques, one might also be able to extend the stability hierarchy in order to adequately cover this case.

The quantifier elimination results in Sect. 4 beg the question to what extent they can be extended to the theories with several derivatives. It would also be interesting to consider how quantifier elimination results for other theories of Boolean algebras, that might require additional predicates, can be extended to quantifier elimination results for the corresponding theory T_1^K. One example is the theory of infinite atomic Boolean algebras, which admits quantifier elimination if one adds predicates for "n atoms lie below x" (see [3] for details and further examples).

References

1. Chatzidakis, Z.: Model theory of difference fields. In: The Notre Dame lectures, vol. 18. Lecture of Notes Logic Association Symbol. Logic, Urbana, pp. 45–96 (2005)
2. Cherlin, G.L., Hrushovski, E.: Finite Structures with Few types, vol. 152. Annals of Mathematics Studies, Princeton (2003)
3. Derakhshan, J., Macintyre, A.: Enrichments of Boolean algebras by Presburger predicates. Fund. Math. **239**(1), 1–17 (2017). ISSN: 0016-2736
4. Esparza, J., Michaux, C., Steinhorn, C. (eds.): Finite and algorithmic model theory. In: London Mathematical Society Lecture Note Series, vol. 379. Cambridge University, Cambridge, pp. xii+341 (2011). ISBN: 978-0-521-71820-2
5. Hodges, W.: Model theory. In: Encyclopedia of Mathematics and its Applications, vol. 42. Cambridge University, Cambridge, pp. xiv+772 (1993). ISBN: 0-521-30442-3
6. Kühnrich, M.: Differentialoperatoren über Booleschen Algebren. Z. Math. Logik Grundlag. Math. **32**(3), 271–288 (1986). ISSN: 0044-3050
7. Macpherson, D., Steinhorn, C.: Definability in classes of finite structures. In: Finite and algorithmic model theory, vol. 379. London Mathematical Society Lecture Note Series. Cambridge University, Cambridge, pp. 140–176 (2011)

8. Marker, D.:Model theory. In: Graduate Texts in Mathematics. An introduction, vol. 217. Springer, New York, pp. viii+342 (2002). ISBN: 0-387-98760-6
9. Pestov, V.: Dynamics of infinite-dimensional groups. In: University Lecture Series, vol. 40. American Mathematical Society, Providence, pp. viii+192 (2006). ISBN: 978-0-8218-4137-2; 0-8218-4137-8
10. Rudeanu, S.: Boolean Functions and Equations. North Holland Publishing Company, Amsterdam (1974)
11. Rudeanu, S.: Lattice Function and Equations. Springer, Berlin (2001)
12. Steinbach, B., Posthoff, C.: Boolean differential calculus. In: Syn- thesis Lectures on Digital Circuits and Systems, vol. 52. Morgan and Claypool, New York (2017)
13. Thayse, A.: Boolean calculus of differences. In: Akers, S.B. (eds.) Lecture Notes in Computer Science, vol. 101. Springer, Berlin, pp. vii+144 (1981). ISBN: 3-540-10286-8

Construction of Binary Bent Functions by FFT-Like Permutation Algorithms

Radomir S. Stanković, Milena Stanković, Claudio Moraga, and Jaakko Astola

1 Introduction

Binary bent functions are Boolean functions which achieve maximal non-linearity of $2^{n-1} - 2^{n/2-1}$, where n is the number of variables and has to be an even natural number. It means that bent functions are at the highest possible distance from all affine functions. Recall that the affine functions are defined as linear functions and their complements. An important characteristic of bent functions is that they must have exactly specified Hamming weight, i.e., the number of non-zero values in the truth-vector, which can be either $w_1 = 2^{n-1} - 2^{n/2-1}$ or $w_2 = 2^{n-1} + 2^{n/2-1}$. In that respect, the set of all bent functions for a given n can be split into two subsets S_{w_1} and S_{w_2} of equal cardinality consisting of functions with the same number w_1 or w_2 of non-zero values. Each function in S_{w_1} has a counterpart in S_{w_2}, which is its logic complement, and vice versa.

Another characteristic of bent functions is their degree d defined as the largest number of variables in a product term in their functional expressions. Under the term functional expression for a binary function, we assume the positive-polarity Reed-Muller expressions [13, 14], initially introduced as Zhegalkin polynomials [18, 19], which in cryptographic and related communities are often called the

R. S. Stanković (✉)
Mathematical Institute of SASA, Belgrade, Serbia

M. Stanković
Department of Computer Science, Faculty of Electronic Engineering, Niš, Serbia

C. Moraga
Faculty of Computer Science, Technical University of Dortmund, Dortmund, Germany

Department of Informatics, Technical University "Federico Santa María", Valparaíso, Chile

J. Astola
Department of Signal Processing, Tampere University of Technology, Tampere, Finland

R. Drechsler, D. Große (eds.), *Recent Findings in Boolean Techniques*,
https://doi.org/10.1007/978-3-030-68071-8_5

algebraic normal form (ANF) [8, 17]. Fast computing algorithms for determining Reed-Muller coefficients corresponding directly to the Cooley-Tukey algorithms for the fast Fourier transform (FFT), which is a fast algorithm for computing the discrete Fourier transform (DFT) [7], have been proposed by Ph. W. Besslich [3–6]. It is usually assumed that the product terms, equivalently, the Reed-Muller coefficients, are arranged in the Hadamard order. The value of the degree d of a bent function determines possible non-zero Reed-Muller coefficients, since it defines product terms to which they can be assigned.

In spectral domain, bent functions are defined in terms of a specific requirement that should be satisfied by the spectral coefficients of a bent function with respect to the discrete Walsh transform that is the Fourier transform on the group on which Boolean functions are defined. The discrete Walsh transform can be computed by the fast Walsh transform (FWT) directly derived from FFT [1, 2]. This is a feature that will be used in this chapter where we refer to a particular FWT corresponding to the Cooley-Tukey FFT based on the Good-Thomas factorization of the DFT transform matrix.

A classical theorem in the theory of bent functions states that functions derived from a bent function by adding to it affine functions are also bent [8]. In other words, adding a linear combination of variables and the constant 1 to a binary bent function preserves the bentness [16]. From the spectral transform point of view, preserving the bentness is viewed as performing the so-called spectral invariant operations meaning that these operations preserve the absolute values of Walsh spectral coefficients. In the spectral domain, these operations are performed as particular permutations of strictly determined subsets of Walsh coefficients and possibly change of their signs. This feature will be used later.

For a given bent function, its non-zero Reed-Muller coefficients uniquely determine in terms of which variables spectral invariant operations, in particular, spectral translation and disjoint spectral translation defined bellow, can be interpreted as adding linear terms to produce new bent functions. For example, if the non-zero Reed-Muller coefficient is assigned to the product term $x_i x_j$, the substitution $x_i \rightarrow x_i \oplus x_j$, which is a spectral invariant operation, will produce a new function with the terms $x_i x_j \oplus x_j$. Thus, it can be viewed as adding the variable x_j to the initial function, and the new function is also bent. In the same way, for such a function, the substitution $x_j \rightarrow x_j \oplus x_i$ can be viewed as adding the variable x_i. In the spectral domain, both these substitutions are performed as particular permutations of precisely defined subsets of Walsh coefficients.

In this chapter, we point out the following.

1. Bent functions of the same Hamming weight are mutually related by permutations of their function values.
2. Since bentness should be preserved, meaning that spectra of bent functions should remain flat, allowed permutations correspond to spectral invariant operations.
3. In the spectral domain, spectral invariant operations require permutations of certain precisely determined subsets of Walsh spectral coefficients.

4. In the functional domain, permutations of spectral coefficients can be equivalently expressed as permutations of function values involved in computing the spectral coefficients to be permuted.
5. When spectral coefficients are computed by the FWT, the function values which should be permuted due to spectral invariant operations are directly and uniquely determined by the factor matrices describing the corresponding steps of the FWT.
6. The structure of the factor matrices in the FWT determines the structure of the permutation matrices mutually relating bent functions with the same Hamming weight.

This leads to an algorithm for constructing bent functions by permutation matrices, which we call the FFT-like in the general case, or FWT-like permutation matrices in the binary case.

In what follows, we first briefly present the basic concepts that will be used, and then introduce the FWT-like matrices and explain their application to construct bent functions.

2 Walsh Transform

The Walsh transform is the Fourier transform on the finite dyadic group which is the natural domain for Boolean functions. Recall that this is a group consisting of binary n-tuples and the group operation is the addition modulo 2. In matrix notation, The Walsh transform can be defined by the transform matrix

$$\mathbf{W}(n) = \bigotimes_{i=1}^{n} \mathbf{W}(1), \quad \mathbf{W}(1) = \begin{bmatrix} 1 & 1 \\ 1 & -1 \end{bmatrix},$$

where $\otimes$ denotes the Kronecker product. This product determines the so-called Hadamard ordering of Walsh functions used in the transform. Emphasizing their ordering here is important, since it determines the position of spectral coefficients assigned to particular Walsh functions within the vector representing the Walsh spectrum $\mathbf{S}_f = [S_f(0), S_f(1), \ldots, S_f(2^n - 1)]^T$ determined for a function $f(x_1, x_2, \ldots, x_n)$ specified by the function vector $\mathbf{F} = [f(0), f(1), \ldots, f(2^n - 1)]^T$ as

$$\mathbf{S}_f = \mathbf{W}(n)\mathbf{F}(n).$$

This feature will be used later when we refer to particular subsets of Walsh coefficients.

When the Walsh transform is applied to a Boolean function, the encoding $(0, 1) \rightarrow (1, -1)$ is usually applied to the function values. In this case, the Walsh coefficients, i.e., elements of the vector representing Walsh spectrum, are even integers between -2^n and 2^n, where n is the number of variables. Further, not all possible combinations of integers in this range are allowed as values of Walsh coefficients of Boolean functions. Certain restrictions are imposed in order to ensure that the inverse transform will produce a Boolean function out of the Walsh spectrum.

3 Spectral Invariant Operations in the Walsh Domain

Any operation over the function values which preserves absolute values of spectral coefficients can be viewed as a spectral invariant operation for a particular spectral transform. In spectral techniques based on the discrete Walsh transform for processing Boolean functions, there are five classical spectral invariant operations, which in the terminology introduced in [11], also used in [12] and elsewhere, are defined as follows:

1. Polarization of the function f
   $f(x_1, x_2, \ldots, x_n) \rightarrow g(x_1, x_2, \ldots, x_n) = f(x_1, x_2, \ldots, x_n) \oplus 1$, which is the logical complement of $f x_1, x_2, \ldots, x_n)$.
2. Polarization of an input variable x_i
   $x_i \rightarrow x_i \oplus 1, i = 1, 2, \ldots, n$.
3. Adding a variable to the function f
   $f(x_1, x_2, \ldots, x_n) \rightarrow g(x_1, x_2, \ldots, x_n) = f(x_1, x_2, \ldots, x_n) \oplus x_i$.
4. Permutation of input variables, $x_i \leftrightarrow x_j$. Thus,

$$f(x_1, \ldots, x_i, \ldots x_j, \ldots, x_n) \rightarrow g(x_1, x_2, \ldots, x_n) = f(x_1, \ldots, x_j, \ldots x_i, \ldots, x_n).$$

5. Substitution of an input variable by a sum of variables containing the replaced variable $x_i \rightarrow x_i \oplus x_r$, i.e.,

$$f(x_1, \ldots, x_i, \ldots, x_n) \rightarrow g(x_1, x_2, \ldots, x_n) = f(x_1, \ldots, x_i \oplus x_r, \ldots, x_n).$$

   Restriction is that the replacing sum must contain the replaced variable x_i.

The operations 3 and 5 are called the disjoint spectral translation and spectral translation, respectively.

In the spectral domain, if indices of spectral coefficients are written in their binary representations, spectral invariant operations can be expressed in the following way:

1. Multiplication of spectral coefficients with -1.
2. Componentwise multiplication of the spectrum with the truth-vector of x_i.
3. Permutation of subsets of spectral coefficients with the index i of the variable x_i added to f,

$$S_{g l_1, \ldots, l_i \oplus 1, \ldots, l_n} \leftrightarrow S_{f l_1, \ldots, l_i, \ldots, l_n},$$

where l_i are binary representations of indices of spectral coefficients. Thus, the spectral coefficients with complemented values of the ith coordinate in the binary representation of indices are permuted.

4. Permutation of subsets of spectral coefficients

$$S_{f l_1,\ldots,l_i,\ldots l_r,\ldots,l_n} \leftrightarrow S_{f l_1,\ldots,l_r,\ldots l_i,\ldots,l_n}.$$

5. Interchange of pairs of spectral coefficients

$$S_{g l_1,\ldots,l_i,l_r \oplus l_i,\ldots,l_n} \leftrightarrow S_{f l_1,\ldots,l_i,l_r,\ldots,l_n}.$$

From the Hadamard ordering of Walsh functions, determined by the Kronecker product structure of the Walsh transform matrix, it follows that subsets of 2^k, $k = 1, 2, \ldots n/2$, Walsh coefficients can be simultaneously permuted or their signs changed, or both, depending on the performed spectral invariant transformations. These subsets of 2^k elements result in a block structure of permutation matrices expressing the permutations due to spectral invariant operations.

4 Bent Functions and Walsh Transform

In the $(1, -1)$ encoding, the sum of values of all Walsh coefficients of a Boolean function is either -2^n or 2^n. Thus, the maximal absolute value of a Walsh coefficient is 2^n in which case all other Walsh coefficients are equal to 0. For example, this is the case for the Walsh coefficient of the index 0, $S_f(0)$ for the constant function 1. The Walsh spectra of Walsh functions exhibit the same property due to the orthogonality of this transform. It means that the ith Walsh coefficient has the value 2^n for the ith Walsh function, while all other coefficients are 0. Recall that the Walsh functions can be viewed as $(1, -1)$ encoding of all possible linear Boolean functions for a given number of variables. For other Boolean functions, the Walsh coefficients are certain combinations of even integers and 0 under the above-mentioned restriction to their sum.

The Walsh coefficients reach the smallest value of the maximum absolute value when all the values of squares of Walsh spectral coefficients are equal mutually [17]. In this case, the Walsh spectrum is flat, and this feature is used as an alternative definition of bent functions. In other words, a Boolean function is bent if the absolute values of all its coefficients $S_f(w)$, $w = 0, 1, \ldots, 2^n - 1$ are $|S_f(w)| = 2^{n/2}$. This illustrates the difference with respect to the linear functions, where a single coefficient takes the maximal absolute value and all others are 0. For bent functions, which are most non-linear functions, all the Walsh coefficients have equal values, and this is the smallest value of the maximum absolute value in the Walsh spectrum of Boolean functions [17].

Spectral invariant operations preserve the flatness of the Walsh spectrum and therefore can be used to derive new bent functions from a given bent function by manipulating with its Walsh spectrum. Notice that spectral invariant operations enumerated above as items 1 and 3, i.e., polarization of the function, and spectral translation actually perform operations allowed by the theorem about adding affine functions to bent functions. Permutation of variables is also not very interesting, since in constructing new bent functions we can start from any bent function, and it is in general irrelevant if we start from a function f or another function derived from it by some permutation of variables. Therefore, the most interesting for considerations in the present considerations are polarization of variables defined as $x_i \rightarrow x_i \oplus 1$, and disjoint spectral translation of variables $x_i \rightarrow x_i \oplus x_j$. From the implementation point of view, the question arises how they can be efficiently implemented, since dealing with functional expressions is impractical especially in the case of functions with a large number of product terms. In this context, establishing links to fast computing algorithms for spectral transforms has sense, since besides being efficient, such algorithms can be performed over either vectors or decision diagrams as underlying data structures to represent functions to be processed. In what follows, we show that these operations can be performed by using permutation matrices derived from fast computing algorithms for the discrete Walsh transform in Hadamard ordering.

5 Essence of FFT

For the completeness of the presentation, in this section we briefly recall the basic ideas of a fast computing algorithm to which we refer in defining permutations used for constructing bent functions.

The fast Fourier transform (FFT) is an algorithm to compute the discrete Fourier transform (DFT). In this chapter, we refer to the so-called Cooley-Tukey FFT that is based on the Good-Thomas decomposition of the DFT transform matrix [7, 9], [10], [15]. The reason for the selection is that this algorithm is usually applied when computing the discrete Walsh transform which is viewed as the Fourier transform on the finite dyadic group of order 2^n, C_2^n, where $C_2 = (\{0, 1\}, \oplus)$, and the group operation is the addition modulo 2. The main idea behind this algorithm, which is called the fast Walsh transform (FWT) [1, 2], is to convert computing a 2^n length transform into performing n transforms of length 2. This means that the transform is performed in n steps, with computations in each step determined by the (2×2) basic Walsh transform matrix $\mathbf{W}(1)$. The $(2^n \times 2^n)$ Walsh matrix $\mathbf{W}(n)$ is decomposed into n factor matrices

$$\mathbf{W}(n) = \prod_{i=1}^{n} \mathbf{C}_i(n), \quad \mathbf{C}_i(n) = \bigotimes_{j=1}^{n} \mathbf{D}_j,$$

where

$$\mathbf{D}_j = \begin{cases} \mathbf{W}(1) & j = i, \\ \mathbf{I}(1), & j \neq i, \end{cases} \quad \mathbf{W}(1) = \begin{bmatrix} 1 & 1 \\ 1 & -1 \end{bmatrix}, \quad \mathbf{I}(1) = \begin{bmatrix} 1 & 0 \\ 0 & 1 \end{bmatrix}.$$

The factor matrices are sparse, and it is known that FFT is the algorithm with the smallest number of operations required to compute a DFT. From there comes the efficiency of FFT.

6 Permutation Matrices

Spectral invariant operations perform permutation of subsets of Walsh coefficients, which in the original domain corresponds to a permutation of function values involved in computing these spectral coefficients to be permuted. These values are determined by the addresses of locations from which the data are fetched in FFT to compute the spectral coefficients. These addresses are determined by the flow-graphs of steps of FFT, equivalently, the factor matrices describing the steps. By referring to them, we define the permutation matrices, which we call FFT-like permutation matrices, to construct from a given bent function other bent functions with the same number of non-zero values. Since spectral invariant operations preserve flat spectra, it follows that permutation matrices in the original domain related to the spectral invariant operations also preserve bentness. Thus, functions constructed by the application of these permutation matrices do not need to be checked for bentness which is a good feature of the approach proposed.

Different combinations of these FFT-like matrices will produce different new bent functions. Further, their application in different order will also produce different new bent functions. A successive application of the same FFT-like permutation matrix twice does not have sense, since they are involutions, i.e., $\mathbf{P}^2(n) = \mathbf{I}(n)$.

Recall that a permutation can be expressed in terms of other permutations in different ways. Therefore, it should be noticed that some combinations of permutation matrices as well as some combinations of the order in which they are applied might produce identical bent functions. This can be viewed as a disadvantage of the proposed method on the one hand; however, on the other hand, this means that a small library of FFT-like permutation matrices is sufficient to produce from a given initial function a considerable number of new bent functions.

We define two basic matrices

$$\mathbf{P}(1) = \begin{bmatrix} 0 & 1 \\ 1 & 0 \end{bmatrix}, \quad \mathbf{I}(1) = \begin{bmatrix} 1 & 0 \\ 0 & 1 \end{bmatrix}.$$

By using these basic matrices, we define $(2^n \times 2^n)$ permutation matrices with respect to the ith variable in a function in n variables as

$$\mathbf{P}_i(n) = \bigotimes_{j=1}^{n} \mathbf{D}_j, \quad \mathbf{D}_j = \begin{cases} \mathbf{P}(1), \ j = i, \\ \mathbf{I}(1), \ \text{otherwise.} \end{cases}$$

It can be observed a strong resemblance in the structure of this matrix $\mathbf{P}_i(n)$ to the factor matrices $\mathbf{C}_i(n)$ in FFT, which gives a justification for the term FFT-like permutation matrix for $\mathbf{P}_i(n)$. This resemblance consists in the replacement of the basic transform matrix $\mathbf{W}(1)$ by the basic permutation matrix $\mathbf{P}(1)$.

From the definition of the spectral invariant operation of polarization of a variable $x_i \rightarrow x_i \oplus 1$ by referring to its expression in the spectral domain, it follows that it is performed by application of the permutation matrix $\mathbf{P}_i$ to the truth-vector $\mathbf{F}$ of a given function f. A recursive application of the permutation matrices assigned to different variables will produce various bent functions. Notice that the recursive application of permutation matrices $\mathbf{P}_i$ and $\mathbf{P}_j$ assigned to the variables x_i and x_j, respectively, to perform their polarization, is equivalent to the multiplication with a permutation matrix $\mathbf{P}_{i,j}$ obtained as the product of the corresponding permutation matrices assigned to variables. Example 1 in Sect. 7 illustrates this procedure over FFT-like permutation matrices.

Now, we define permutation matrices involving basic matrices at the positions of two variables in steps of FFT. To determine these matrices, we use auxiliary symbolic matrices as follows.

Consider two symbolic (2×2) matrices $\mathbf{L}(1) = \begin{bmatrix} a & b \\ b & a \end{bmatrix}$ and $\mathbf{T}(1) = \begin{bmatrix} a & 0 \\ 0 & b \end{bmatrix}$. An $(2^n \times 2^n)$ auxiliary symbolic matrix is defined as

$$\mathbf{A}_{i,k}(n) = \bigotimes_{j=1}^{n} \mathbf{A}_j, \quad \mathbf{A}_j = \begin{cases} \mathbf{L}(1), \ j = i, \\ \mathbf{T}(1), \ j = k, \\ \mathbf{I}(1), \ \text{otherwise.} \end{cases}$$

The permutation matrix $\mathbf{Q}_{i,k}(n)$ is defined as a matrix derived from $\mathbf{A}_{i,k}(n)$ by replacement of symbols a^2 and b^2 by 1, while symbols with mixed letters ab and ba are replaced by 0. Again, from definition of spectral invariant operations in the spectral domain, if follows that the permutation matrix $\mathbf{Q}_{i,k}(n)$ performs the substitution of a variable x_i by $x_i \oplus x_k$. Example 2 in Sect. 7 illustrates the application of this permutation matrix.

7 Illustrative Examples

In this section, we present examples which illustrate the definition and the application of FFT-like permutation matrices $\mathbf{P}_i(n)$, $\mathbf{P}_{i,j}(n)$, and $\mathbf{Q}_{i,k}(n)$.

Example 1 Consider the function $f(x_1, x_2, x_3, x_4) = x_1x_2 \oplus x_3x_4 \oplus x_1x_3 \oplus x_2$. Its truth-vector is

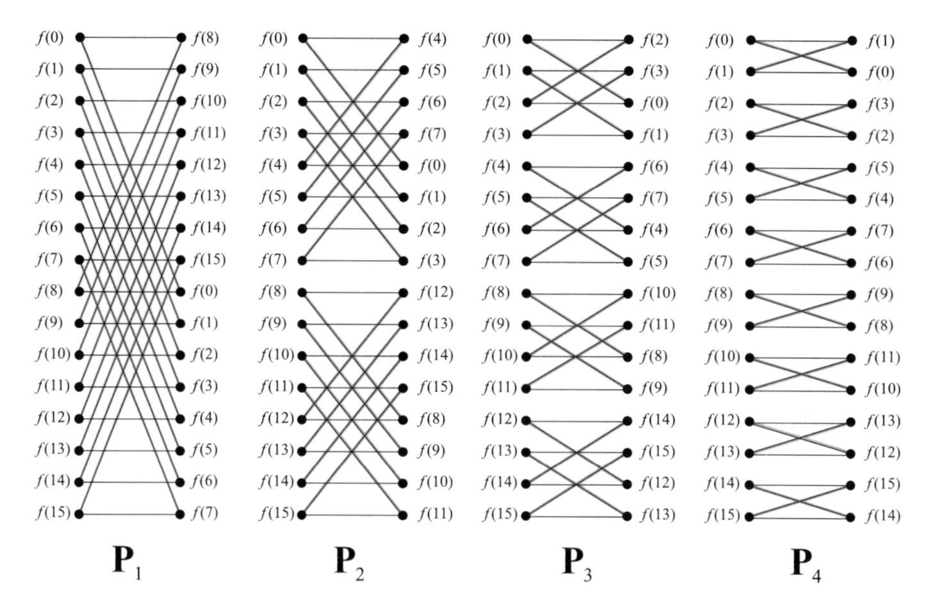

Fig. 1 Flow-graph for computing with $\mathbf{P}_1, \mathbf{P}_2, \mathbf{P}_3, \mathbf{P}_4$

$$\mathbf{F} = [0, 0, 0, 1, 1, 1, 1, 0, 0, 0, 1, 0, 0, 0, 1, 0]^T.$$

The Walsh spectrum after encoding $(0, 1) \to (1, -1)$ is computed as

$$\mathbf{S}_f = [4, -4, 4, 4, 4, 4, 4, -4, -4, 4, -4, -4, 4, 4, 4, -4]^T.$$

We apply different permutation matrices derived from steps of the fast Walsh transform and produce 15 different bent functions. For each permutation matrix, we show the truth-vector of the produced function, the Walsh spectrum computed in the $(0, 1) \to (1, -1)$ encoding of Boolean values, and the functional expression for the new bent function produced in this way. Table 1 and its continuation Table 2 show the performed spectral invariant operations, the corresponding permutation matrices used in computing, the truth-vectors of the produced new bent functions, their Walsh spectra, and functional expressions. Figure 1 shows the flow-graph for computing with permutation matrices $\mathbf{P}_1$, $\mathbf{P}_2$, $\mathbf{P}_3$, and $\mathbf{P}_4$. The black lines corresponds to the flow-graph of the FWT, while thicker green lines over them show the performed permutations. Figures 2 and 3 show the flow-graph for computing with the matrices $\mathbf{Q}_4$ and $\mathbf{R}_5$. We see that the steps of the FWT are selected depending on the position of the basic permutation matrices in the Kronecker product defining these matrices.

The following example illustrates the construction of a new bent function by application of the permutation matrix corresponding to the spectral invariant operation called the disjoint translation $x_i \to x_i \oplus x_j$.

Table 1 Spectral invariant operations, permutation matrices, truth-vectors of new functions, their Walsh spectra, and functional expressions for produced bent functions from the function f in Example 1

Case 1	$x_1 \to x_1 \oplus 1$
	$\mathbf{P}_1 = \mathbf{P}(1) \otimes \mathbf{I}(1) \otimes \mathbf{I}(1) \otimes \mathbf{I}(1)$
	$\mathbf{F}_1 = \mathbf{P}_1\mathbf{F} = [0, 0, 1, 0, 0, 0, 1, 0, 0, 0, 0, 1, 1, 1, 1, 0]^T$
	$\mathbf{S}_{f1} = [4, -4, 4, 4, 4, 4, 4, -4, 4, -4, 4, 4, -4, -4, -4, 4]^T$
	$f_1 = x_3 \oplus x_3x_4 \oplus x_1x_3 \oplus x_1x_2$
Case 2	$x_2 \to x_2 \oplus 1$
	$\mathbf{P}_2 = \mathbf{I}(1) \otimes \mathbf{P}(1) \otimes \mathbf{I}(1) \otimes \mathbf{I}(1)$
	$\mathbf{F}_2 = \mathbf{P}_2\mathbf{F} = [1, 1, 1, 0, 0, 0, 0, 1, 0, 0, 1, 0, 0, 0, 1, 0]^T$
	$\mathbf{S}_{f2} = [4, -4, 4, 4, -4, -4, -4, 4, -4, 4, -4, -4, -4, -4, -4, 4]^T$
	$f_2 = 1 \oplus x_3x_4 \oplus x_1 \oplus x_1 \oplus x_1x_3 \oplus x_1x_2$
Case 3	$x_3 \to x_3 \oplus 1$
	$\mathbf{P}_3 = \mathbf{I}(1) \otimes \mathbf{I}(1) \otimes \mathbf{P}(1) \otimes \mathbf{I}(1)$
	$\mathbf{F}_3 = \mathbf{P}_3\mathbf{F} = [0, 1, 0, 0, 1, 0, 1, 1, 1, 0, 0, 0, 1, 0, 0, 0]^T$
	$\mathbf{S}_{f3} = [4, -4, -4, -4, 4, 4, -4, 4, -4, 4, 4, 4, 4, 4, -4, 4]^T$
	$f_3 = x_4 \oplus x_3x_4 \oplus x_2 \oplus x_1 \oplus x_1x_3 \oplus x_1x_2$
Case 4	$x_4 \to x_4 \oplus 1$
	$\mathbf{P}_4 = \mathbf{I}(1) \otimes \mathbf{I}(1) \otimes \mathbf{I}(1) \otimes \mathbf{P}(1)$
	$\mathbf{F}_4 = \mathbf{P}_4\mathbf{F} = [0, 0, 1, 0, 1, 1, 0, 1, 0, 0, 0, 1, 0, 0, 0, 1]^T$
	$\mathbf{S}_{f4} = [4, 4, 4, -4, 4, -4, 4, 4, -4, -4, -4, 4, 4, -4, 4, 4]^T$
	$f_4 = x_3 \oplus x_3x_4 \oplus x_2 \oplus x_1x_3 \oplus x_1x_2$
Case 5	$x_2 \to x_2 \oplus 1$ to $(x_1 \to x_1 \oplus 1)$
	$\mathbf{Q}_1 = \mathbf{P}(1) \otimes \mathbf{P}(1) \otimes \mathbf{I}(1) \otimes \mathbf{I}(1)$
	$\mathbf{F}_5 = \mathbf{Q}_1\mathbf{F} = [0, 0, 1, 0, 0, 0, 1, 0, 1, 1, 1, 0, 0, 0, 0, 1]^T$
	$\mathbf{S}_{f5} = [4, -4, 4, 4, -4, -4, -4, 4, 4, -4, 4, 4, 4, 4, 4, -4]^T$
	$f_5 = x_3 \oplus x_3x_4 \oplus x_1 \oplus x_1x_3 \oplus x_1x_2$
Case 6	$x_3 \to x_3 \oplus 1$ to $(x_2 \to x_2 \oplus 1)$
	$\mathbf{Q}_2 = \mathbf{I}(1) \otimes \mathbf{P}(1) \otimes \mathbf{P}(1) \otimes \mathbf{I}(1)$
	$\mathbf{F}_6 = \mathbf{Q}_2\mathbf{F} = [1, 0, 1, 1, 0, 1, 0, 0, 1, 0, 0, 0, 1, 0, 0, 0]^T$
	$\mathbf{S}_{f6} = [4, -4, -4, -4, -4, -4, 4, -4, -4, 4, 4, 4, -4, -4, 4, -4]^T$
	$f_6 = 1 \oplus x_4 \oplus x_3x_4 \oplus x_2 \oplus x_1x_3 \oplus x_1x_2$
Case 7	$x_4 \to x_4 \oplus 1$ to $(x_3 \to x_3 \oplus 1)$
	$\mathbf{Q}_3 = \mathbf{I}(1) \otimes \mathbf{I}(1) \otimes \mathbf{P}(1) \otimes \mathbf{P}(1)$
	$\mathbf{F}_7 = \mathbf{Q}_3\mathbf{F} = [1, 0, 0, 0, 0, 1, 1, 1, 0, 1, 0, 0, 0, 1, 0, 0]^T$
	$\mathbf{S}_{f7} = [4, 4, -4, 4, 4, -4, -4, -4, -4, -4, 4, -4, 4, -4, -4, -4]^T$
	$f_7 = 1 \oplus x_4 \oplus x_3 \oplus x_3x_4 \oplus x_2 \oplus x_1 \oplus x_1x_3 \oplus x_1x_2$
Case 8	$x_3 \to x_3 \oplus 1$ to $(x_1 \to x_1 \oplus 1)$
	$\mathbf{Q}_4 = \mathbf{P}(1) \otimes \mathbf{I}(1) \otimes \mathbf{P}(1) \otimes \mathbf{I}(1)$
	$\mathbf{F}_8 = \mathbf{Q}_4\mathbf{F} = [1, 0, 0, 0, 1, 0, 0, 0, 0, 1, 0, 0, 1, 0, 1, 1]^T$
	$\mathbf{S}_{f8} = [4, -4, -4, -4, 4, 4, -4, 4, 4, -4, -4, -4, -4, -4, 4, -4]^T$
	$f_8 = 1 \oplus x_4 \oplus x_3 \oplus x_3x_4 \oplus x_1 \oplus x_1x_3 \oplus x_1x_2$

(continued)

Table 1 (continued)

Case 9	$x_4 \rightarrow x_4 \oplus 1$ to $(x_2 \rightarrow x_2 \oplus 1)$
	$\mathbf{Q}_5 = \mathbf{I}(1) \otimes \mathbf{P}(1) \otimes \mathbf{I}(1) \otimes \mathbf{P}(1)$
	$\mathbf{F}_9 = \mathbf{Q}_5\mathbf{F} = [1, 1, 0, 1, 0, 0, 1, 0, 0, 0, 0, 1, 0, 0, 0, 1]^T$
	$\mathbf{S}_{f9} = [4, 4, 4, -4, -4, 4, -4, -4, -4, -4, -4, 4, -4, 4, -4, -4]^T$
	$f_9 = 1 \oplus x_3 \oplus x_3x_4 \oplus x_2 \oplus x_1 \oplus x_1x_3 \oplus x_1x_2$
Case 10	$x_4 \rightarrow x_4 \oplus 1$ to $(x_1 \rightarrow x_1 \oplus 1)$
	$\mathbf{Q}_6 = \mathbf{P}(1) \otimes \mathbf{I}(1) \otimes \mathbf{I}(1) \otimes \mathbf{P}(1)$
	$\mathbf{F}_{10} = \mathbf{Q}_6\mathbf{F} = [0, 0, 0, 1, 0, 0, 0, 1, 0, 0, 1, 0, 1, 1, 0, 1]^T$
	$\mathbf{S}_{f10} = [4, 4, 4, -4, 4, -4, 4, 4, 4, 4, 4, -4, -4, 4, -4, -4]^T$
	$f_{10} = x_3x_4 \oplus x_1x_3 \oplus x_1x_2$

Table 2 Substitution, permutation matrix, truth-vector of new function f, its Walsh spectrum, and functional expression for it

Case 11	$x_3 \rightarrow x_3 \oplus 1$ to $(x_2 \rightarrow x_2 \oplus 1)$ to $(x_1 \rightarrow x_1 \oplus 1)$
	$\mathbf{R}_1 = \mathbf{P}(1) \otimes \mathbf{P}(1) \otimes \mathbf{P}(1) \otimes \mathbf{I}(1)$
	$\mathbf{F}_{11} = \mathbf{R}_1\mathbf{F} = [1, 0, 0, 0, 1, 0, 0, 0, 1, 0, 1, 1, 0, 1, 0, 0]^T$
	$\mathbf{S}_{f11} = [4, -4, -4, -4, -4, -4, 4, -4, 4, -4, -4, -4, 4, 4, -4, 4]^T$
	$f_{11} = 1 \oplus x_4 \oplus x_3 \oplus x_3x_4 \oplus x_1x_3 \oplus x_1x_2$
Case 12	$x_4 \rightarrow x_4 \oplus 1$ to $(x_2 \rightarrow x_2 \oplus 1)$ to $(x_1 \rightarrow x_1 \oplus 1)$
	$\mathbf{R}_2 = \mathbf{P}(1) \otimes \mathbf{P}(1) \otimes \mathbf{I}(1) \otimes \mathbf{P}(1)$
	$\mathbf{F}_{12} = \mathbf{R}_2\mathbf{F} = [0, 0, 0, 1, 0, 0, 0, 1, 1, 1, 0, 1, 0, 0, 1, 0]^T$
	$\mathbf{S}_{f12} = [4, 4, 4, -4, -4, 4, -4, -4, 4, 4, 4, -4, 4, -4, 4, 4]^T$
	$f_{12} = x_1 \oplus x_3x_4 \oplus x_1x_3 \oplus x_1x_2$
Case 13	$x_4 \rightarrow x_4 \oplus 1$ to $(x_3 \rightarrow x_3 \oplus 1)$ to $(x_1 \rightarrow x_1 \oplus 1)$
	$\mathbf{R}_3 = \mathbf{P}(1) \otimes \mathbf{I}(1) \otimes \mathbf{P}(1) \otimes \mathbf{P}(1)$
	$\mathbf{F}_{13} = \mathbf{R}_3\mathbf{F} = [0, 1, 0, 0, 0, 1, 0, 0, 1, 0, 0, 0, 0, 1, 1, 1]^T$
	$\mathbf{S}_{f13} = [4, 4, -4, 4, 4, -4, -4, -4, 4, 4, -4, 4, -4, 4, 4, 4]^T$
	$f_{13} = x_4 \oplus x_3x_4 \oplus x_1 \oplus x_1x_3 \oplus x_1x_2$
Case 14	$x_4 \rightarrow x_4 \oplus 1$ to $(x_3 \rightarrow x_3 \oplus 1)$ to $(x_2 \rightarrow x_2 \oplus 1)$
	$\mathbf{R}_4 = \mathbf{I}(1) \otimes \mathbf{P}(1) \otimes \mathbf{P}(1) \otimes \mathbf{P}(1)$
	$\mathbf{F}_{14} = \mathbf{R}_4\mathbf{F} = [0, 1, 1, 1, 1, 0, 0, 0, 0, 1, 0, 0, 0, 1, 0, 0]^T$
	$\mathbf{S}_{f14} = [4, 4, -4, 4, -4, 4, 4, 4, -4, -4, 4, -4, -4, 4, 4, 4]^T$
	$f_{14} = x_4 \oplus x_3 \oplus x_2 \oplus x_3x_4 \oplus x_1x_3 \oplus x_1x_2$
Case 15	$x_4 \rightarrow x_4 \oplus 1$ to $(x_3 \rightarrow x_3 \oplus 1)$ to $(x_2 \rightarrow x_2 \oplus 1)$ to $(x_1 \rightarrow x_1 \oplus 1)$
	$\mathbf{R}_5 = \mathbf{P}(1) \otimes \mathbf{P}(1) \otimes \mathbf{P}(1) \otimes \mathbf{P}(1)$
	$\mathbf{F}_{15} = \mathbf{R}_5\mathbf{F} = [0, 1, 0, 0, 0, 1, 0, 0, 0, 1, 1, 1, 1, 0, 0, 0]^T$
	$\mathbf{S}_{f15} = [4, 4, -4, 4, -4, 4, 4, 4, 4, 4, -4, 4, 4, -4, -4, -4]^T$
	$f_{15} = x_4 \oplus x_3x_4 \oplus x_1x_3 \oplus x_1x_2$

Example 2 We determine a symbolic matrix as

$$\mathbf{A}_{1,2} = \mathbf{L}(1) \otimes \mathbf{T}(1) \otimes \mathbf{I}(1) \otimes \mathbf{I}(1).$$

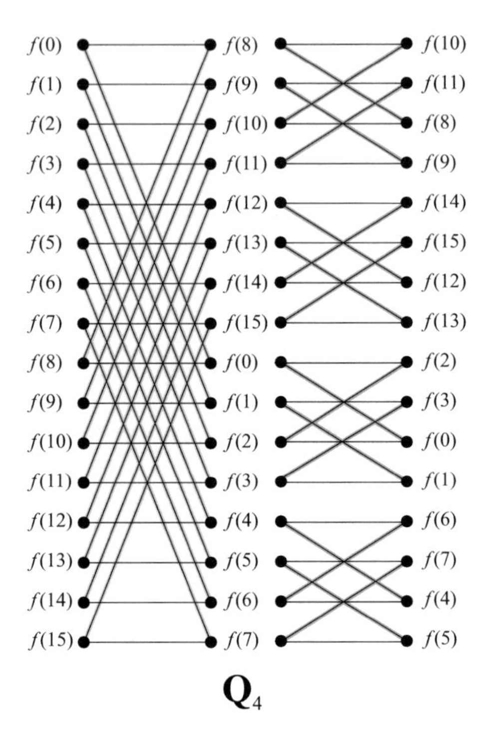

Fig. 2 Flow-graph for computing with $\mathbf{Q}_4$

After performing a symbolic computation and replacement of elements a^2 and b^2 by 1 and all other elements by 0, we get the permutation matrix which can be written in a condensed form as

$$\mathbf{Q}_{1,2} = \left[\begin{array}{cccc|cccc} \mathbf{I}(1) & \mathbf{0}(1) & \mathbf{0}(1) & \mathbf{0}(1) & \mathbf{0}(1) & \mathbf{0}(1) & \mathbf{0}(1) & \mathbf{0}(1) \\ \mathbf{0}(1) & \mathbf{I}(1) & \mathbf{0}(1) & \mathbf{0}(1) & \mathbf{0}(1) & \mathbf{I}(1) & \mathbf{0}(1) & \mathbf{0}(1) \\ \mathbf{0}(1) & \mathbf{0}(1) & \mathbf{I}(1) & \mathbf{0}(1) & \mathbf{0}(1) & \mathbf{0}(1) & \mathbf{I}(1) & \mathbf{0}(1) \\ \mathbf{0}(1) & \mathbf{0}(1) & \mathbf{0}(1) & \mathbf{0}(1) & \mathbf{0}(1) & \mathbf{0}(1) & \mathbf{0}(1) & \mathbf{I}(1) \\ \hline \mathbf{0}(1) & \mathbf{0}(1) & \mathbf{0}(1) & \mathbf{0}(1) & \mathbf{I}(1) & \mathbf{0}(1) & \mathbf{0}(1) & \mathbf{0}(1) \\ \mathbf{0}(1) & \mathbf{I}(1) & \mathbf{0}(1) & \mathbf{0}(1) & \mathbf{0}(1) & \mathbf{I}(1) & \mathbf{0}(1) & \mathbf{0}(1) \\ \mathbf{0}(1) & \mathbf{0}(1) & \mathbf{I}(1) & \mathbf{0}(1) & \mathbf{0}(1) & \mathbf{0}(1) & \mathbf{I}(1) & \mathbf{0}(1) \\ \mathbf{0}(1) & \mathbf{0}(1) & \mathbf{0}(1) & \mathbf{I}(1) & \mathbf{0}(1) & \mathbf{0}(1) & \mathbf{0}(1) & \mathbf{0}(1) \end{array}\right],$$

where $\mathbf{I}(1)$ and $\mathbf{0}(1)$ are the (2×2) identity and the zero matrices.

Figure 4 shows the flow-graph for computing with this matrix.

The matrix $\mathbf{Q}_{1,2}$ applied to the truth-vector $\mathbf{F}$ of the initial function in Example 1 produces the truth vector $\mathbf{F}_{1,2}$ of a function $f_{1,2}(x_1, x_2, x_3, x_4)$ as

$$\mathbf{F}_{1,2} = [0, 0, 0, 1, 0, 0, 1, 0, 0, 0, 1, 0, 1, 1, 1, 0]^T$$
$$\mathbf{S}_{f1,2} = [4, -4, 4, 4, 4, 4, 4, -4, 4, 4, 4, -4, -4, 4, -4, -4]^T$$
$$f_{1,2}(x_1, x_2, x_3, x_4) = x_3x_4 \oplus x_2x_3 \oplus x_1x_3 \oplus x_1x_2.$$

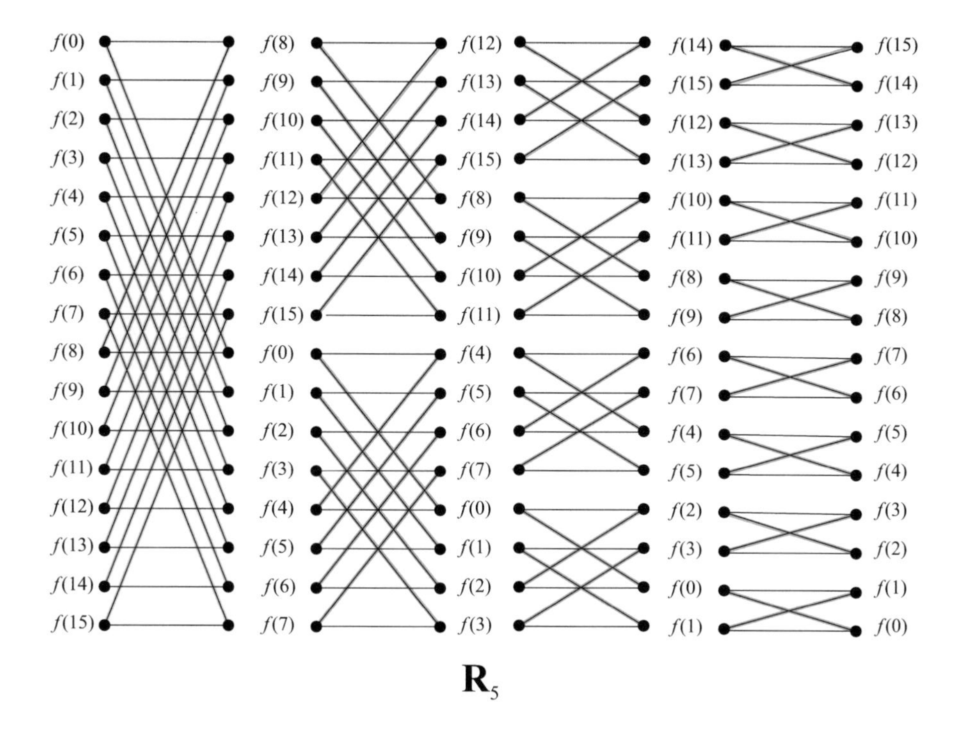

Fig. 3 Flow-graph for computing with $\mathbf{R}_5$

The same function can be derived from f by the substitution $x_1 \rightarrow x_1 \oplus x_2$.

Define a symbolic matrix

$$\mathbf{A}_{1,3} = \mathbf{L}(1) \otimes \mathbf{I}(1) \otimes \mathbf{T}(1) \otimes \mathbf{I}(1).$$

From this matrix, we produce that permutation matrix, which can be written as

$$\mathbf{Q}_{1,3} = \left[\begin{array}{cc|cc|cc|cc}
\mathbf{I}(1) & \mathbf{0}(1) & \mathbf{0}(1) & \mathbf{0}(1) & \mathbf{0}(1) & \mathbf{0}(1) & \mathbf{0}(1) & \mathbf{0}(1) \\
\mathbf{0}(1) & \mathbf{0}(1) & \mathbf{0}(1) & \mathbf{0}(1) & \mathbf{0}(1) & \mathbf{I}(1) & \mathbf{0}(1) & \mathbf{0}(1) \\
\hline
\mathbf{0}(1) & \mathbf{0}(1) & \mathbf{I}(1) & \mathbf{0}(1) & \mathbf{0}(1) & \mathbf{0}(1) & \mathbf{0}(1) & \mathbf{0}(1) \\
\mathbf{0}(1) & \mathbf{0}(1) & \mathbf{0}(1) & \mathbf{0}(1) & \mathbf{0}(1) & \mathbf{0}(1) & \mathbf{0}(1) & \mathbf{I}(1) \\
\hline
\mathbf{0}(1) & \mathbf{0}(1) & \mathbf{0}(1) & \mathbf{0}(1) & \mathbf{I}(1) & \mathbf{0}(1) & \mathbf{0}(1) & \mathbf{0}(1) \\
\mathbf{0}(1) & \mathbf{I}(1) & \mathbf{0}(1) & \mathbf{0}(1) & \mathbf{0}(1) & \mathbf{0}(1) & \mathbf{0}(1) & \mathbf{0}(1) \\
\hline
\mathbf{0}(1) & \mathbf{0}(1) & \mathbf{0}(1) & \mathbf{0}(1) & \mathbf{0}(1) & \mathbf{0}(1) & \mathbf{I}(1) & \mathbf{0}(1) \\
\mathbf{0}(1) & \mathbf{0}(1) & \mathbf{0}(1) & \mathbf{I}(1) & \mathbf{0}(1) & \mathbf{0}(1) & \mathbf{0}(1) & \mathbf{0}(1)
\end{array}\right].$$

Figure 5 shows the flow-graph for computing with this matrix.

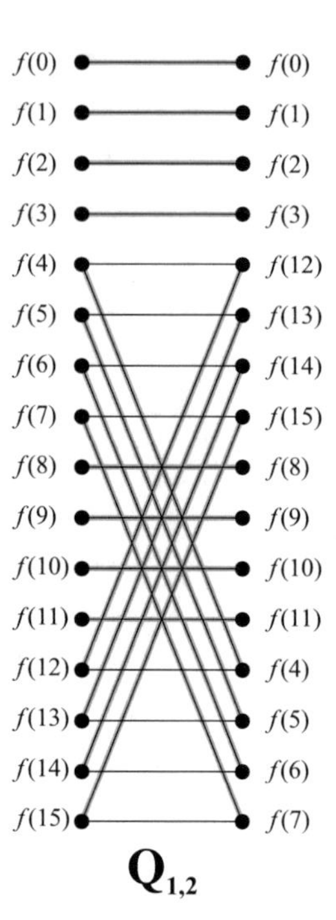

Fig. 4 Flow-graph for computing with $\mathbf{Q}_{1,2}$

Application of this permutation matrix to the truth-vector of the considered initial function f produces the truth-vector of the function $f_{1,3}$ as

$$\mathbf{F}_{1,3} = [0, 0, 1, 0, 1, 1, 1, 0, 0, 0, 0, 1, 0, 0, 1, 0]^T$$
$$\mathbf{S}_{f1,3} = [4, -4, 4, 4, 4, 4, 4, -4, -4, -4, -4, 4, 4, -4, 4, 4]^T$$
$$f_{1,3}(x_1, x_2, x_3, x_4) = x_3 \oplus x_3x_4 \oplus x_2 \oplus x_2x_3 \oplus x_1x_3 \oplus x_1x_2$$

The same function can be obtained from the functional expression for the initial function f by the spectral invariant operation $x_1 \to x_1 \oplus x_3$.

Define a symbolic matrix

$$\mathbf{A}_{1,4} = \mathbf{L}(1) \otimes \mathbf{I}(1) \otimes \mathbf{I}(1) \otimes \mathbf{T}(1),$$

which results in the permutation matrix

$$\mathbf{Q}_{1,4} = \left[\begin{array}{cc|cc} \mathbf{C} & \mathbf{0} & \mathbf{D} & \mathbf{0} \\ \mathbf{0} & \mathbf{C} & \mathbf{0} & \mathbf{D} \\ \hline \mathbf{D} & \mathbf{0} & \mathbf{C} & \mathbf{0} \\ \mathbf{0} & \mathbf{D} & \mathbf{0} & \mathbf{C} \end{array}\right], \quad \text{where} \quad \mathbf{C} = \left[\begin{array}{cc|cc} 1 & 0 & 0 & 0 \\ 0 & 0 & 0 & 0 \\ \hline 0 & 0 & 1 & 0 \\ 0 & 0 & 0 & 0 \end{array}\right], \quad \text{and} \quad \mathbf{D} = \left[\begin{array}{cc|cc} 0 & 0 & 0 & 0 \\ 0 & 1 & 0 & 0 \\ \hline 0 & 0 & 0 & 0 \\ 0 & 0 & 0 & 1 \end{array}\right].$$

Thus,

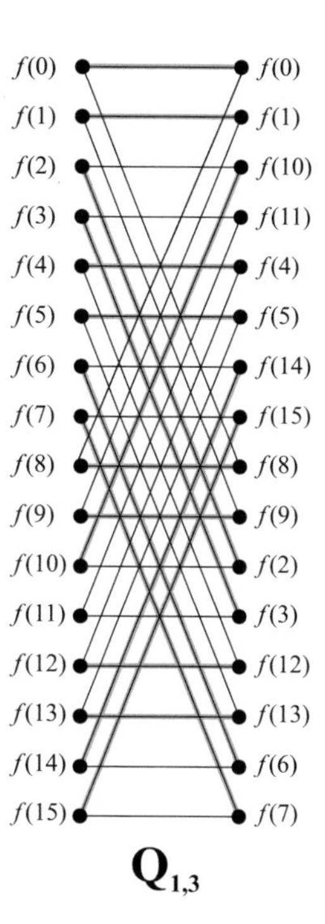

Fig. 5 Flow-graph for computing with $\mathbf{Q}_{1,3}$

$$\mathbf{Q}_{1,4} = \left[\begin{array}{cc|cc|cc|cc}
\mathbf{C}(1) & \mathbf{0}(1) & \mathbf{0}(1) & \mathbf{0}(1) & \mathbf{D}(1) & \mathbf{0}(1) & \mathbf{0}(1) & \mathbf{0}(1) \\
\mathbf{0}(1) & \mathbf{C}(1) & \mathbf{0}(1) & \mathbf{0}(1) & \mathbf{0}(1) & \mathbf{D}(1) & \mathbf{0}(1) & \mathbf{0}(1) \\
\hline
\mathbf{0}(1) & \mathbf{0}(1) & \mathbf{C}(1) & \mathbf{0}(1) & \mathbf{0}(1) & \mathbf{0}(1) & \mathbf{D}(1) & \mathbf{0}(1) \\
\mathbf{0}(1) & \mathbf{0}(1) & \mathbf{0}(1) & \mathbf{C}(1) & \mathbf{0}(1) & \mathbf{0}(1) & \mathbf{0}(1) & \mathbf{D}(1) \\
\hline
\mathbf{D}(1) & \mathbf{0}(1) & \mathbf{0}(1) & \mathbf{0}(1) & \mathbf{C}(1) & \mathbf{0}(1) & \mathbf{0}(1) & \mathbf{0}(1) \\
\mathbf{0}(1) & \mathbf{D}(1) & \mathbf{0}(1) & \mathbf{0}(1) & \mathbf{0}(1) & \mathbf{C}(1) & \mathbf{0}(1) & \mathbf{0}(1) \\
\hline
\mathbf{0}(1) & \mathbf{0}(1) & \mathbf{D}(1) & \mathbf{0}(1) & \mathbf{0}(1) & \mathbf{0}(1) & \mathbf{C}(1) & \mathbf{0}(1) \\
\mathbf{0}(1) & \mathbf{0}(1) & \mathbf{0}(1) & \mathbf{D}(1) & \mathbf{0}(1) & \mathbf{0}(1) & \mathbf{0}(1) & \mathbf{C}(1)
\end{array}\right],$$

Figure 6 shows the flow-graph for computing with this matrix.
This matrix produces a new bent function $f_{1,4}$ specified as

$$\mathbf{F}_{1,4} = [0, 0, 0, 0, 1, 0, 1, 0, 0, 0, 1, 1, 0, 1, 1, 0]^T$$
$$\mathbf{S}_{f1,4} = [4, -4, 4, 4, 4, 4, 4, -4, 4, -4, -4, -4, 4, 4, -4, 4]^T$$
$$f_{1,4}(x_1, x_2, x_3, x_4) = x_2 \oplus x_2x_4 \oplus x_1x_3 \oplus x_1x_2$$

The same function can be obtained by the substitution of the variable $x_1 \rightarrow x_1 \oplus x_4$.

Define a symbolic matrix

$$\mathbf{A}_{2,3} = \mathbf{I}(1) \otimes \mathbf{L}(1) \otimes \mathbf{T}(1) \otimes \mathbf{I}(1),$$

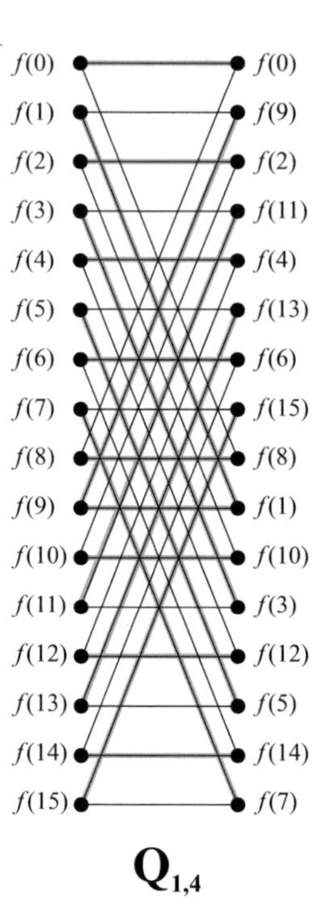

Fig. 6 Flow-graph for computing with $\mathbf{Q}_{1,4}$

which produces the permutation matrix

$$\mathbf{Q}_{2,3} = \left[\begin{array}{cc|cc|cc|cc} \mathbf{I}(1) & \mathbf{0}(1) & \mathbf{0}(1) & \mathbf{0}(1) & \mathbf{0}(1) & \mathbf{0}(1) & \mathbf{0}(1) & \mathbf{0}(1) \\ \mathbf{0}(1) & \mathbf{0}(1) & \mathbf{0}(1) & \mathbf{I}(1) & \mathbf{0}(1) & \mathbf{0}(1) & \mathbf{0}(1) & \mathbf{0}(1) \\ \hline \mathbf{0}(1) & \mathbf{0}(1) & \mathbf{I}(1) & \mathbf{0}(1) & \mathbf{0}(1) & \mathbf{0}(1) & \mathbf{0}(1) & \mathbf{0}(1) \\ \mathbf{0}(1) & \mathbf{I}(1) & \mathbf{0}(1) & \mathbf{0}(1) & \mathbf{0}(1) & \mathbf{0}(1) & \mathbf{0}(1) & \mathbf{0}(1) \\ \hline \mathbf{0}(1) & \mathbf{0}(1) & \mathbf{0}(1) & \mathbf{0}(1) & \mathbf{I}(1) & \mathbf{0}(1) & \mathbf{0}(1) & \mathbf{0}(1) \\ \mathbf{0}(1) & \mathbf{0}(1) & \mathbf{0}(1) & \mathbf{0}(1) & \mathbf{0}(1) & \mathbf{0}(1) & \mathbf{0}(1) & \mathbf{I}(1) \\ \hline \mathbf{0}(1) & \mathbf{0}(1) & \mathbf{0}(1) & \mathbf{0}(1) & \mathbf{0}(1) & \mathbf{0}(1) & \mathbf{I}(1) & \mathbf{0}(1) \\ \mathbf{0}(1) & \mathbf{0}(1) & \mathbf{0}(1) & \mathbf{0}(1) & \mathbf{0}(1) & \mathbf{I}(1) & \mathbf{0}(1) & \mathbf{0}(1) \end{array}\right].$$

Figure 7 shows the flow-graph for computing with this matrix.

This matrix converts the initial function f into a function $f_{2,3}$ specified by the truth-vector.

$$\begin{aligned} \mathbf{F}_{2,3} &= [0, 0, 1, 0, 1, 1, 0, 1, 0, 0, 1, 0, 0, 0, 1, 0]^T \\ \mathbf{S}_{f2,3} &= [4, -4, 4, 4, 4, -4, 4, 4, -4, 4, -4, -4, 4, -4, 4, 4]^T \\ f_{2,3}(x_1, x_2, x_3, x_4) &= x_3 \oplus x_3x_4 \oplus x_2 \oplus x_1x_2 \end{aligned}$$

The same function can be obtained by the substitution of variable $x_2 \rightarrow x_2 \oplus x_3$.

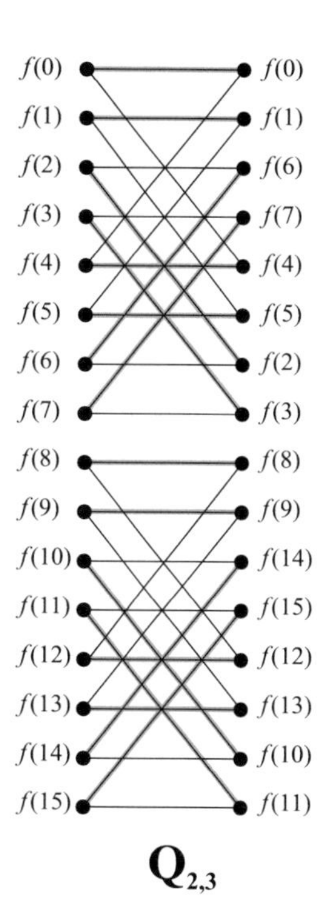

Fig. 7 Flow-graph for computing with $\mathbf{Q}_{2,3}$

Consider a matrix obtained as the product of two permutation matrices performing disjoint spectral translation with respect to two different variables

$$\mathbf{Q}_{1,2,2,3} = \mathbf{Q}_{1,2} \cdot \mathbf{Q}_{2,3}.$$

It can be written in condensed notation as

$$\mathbf{Q}_{1,2,2,3} = \left[\begin{array}{cccc|cccc} \mathbf{I}(1) & \mathbf{0}(1) & \mathbf{0}(1) & \mathbf{0}(1) & \mathbf{0}(1) & \mathbf{0}(1) & \mathbf{0}(1) & \mathbf{0}(1) \\ \mathbf{0}(1) & \mathbf{0}(1) & \mathbf{0}(1) & \mathbf{I}(1) & \mathbf{0}(1) & \mathbf{0}(1) & \mathbf{0}(1) & \mathbf{0}(1) \\ \mathbf{0}(1) & \mathbf{0}(1) & \mathbf{0}(1) & \mathbf{0}(1) & \mathbf{0}(1) & \mathbf{0}(1) & \mathbf{I}(1) & \mathbf{0}(1) \\ \mathbf{0}(1) & \mathbf{0}(1) & \mathbf{0}(1) & \mathbf{0}(1) & \mathbf{0}(1) & \mathbf{I}(1) & \mathbf{0}(1) & \mathbf{0}(1) \\ \hline \mathbf{0}(1) & \mathbf{0}(1) & \mathbf{0}(1) & \mathbf{0}(1) & \mathbf{I}(1) & \mathbf{0}(1) & \mathbf{0}(1) & \mathbf{0}(1) \\ \mathbf{0}(1) & \mathbf{0}(1) & \mathbf{0}(1) & \mathbf{0}(1) & \mathbf{0}(1) & \mathbf{0}(1) & \mathbf{0}(1) & \mathbf{I}(1) \\ \mathbf{0}(1) & \mathbf{0}(1) & \mathbf{I}(1) & \mathbf{0}(1) & \mathbf{0}(1) & \mathbf{0}(1) & \mathbf{0}(1) & \mathbf{0}(1) \\ \mathbf{0}(1) & \mathbf{I}(1) & \mathbf{0}(1) & \mathbf{0}(1) & \mathbf{0}(1) & \mathbf{0}(1) & \mathbf{0}(1) & \mathbf{0}(1) \end{array}\right].$$

Figure 8 shows the flow-graph for computing with this matrix.

The application of this matrix to the initial function f produces a function with the truth-vector

$$\mathbf{F}_{1,2,2,3} = [0, 0, 1, 0, 0, 0, 1, 0, 0, 0, 1, 0, 1, 1, 0, 1]^T$$
$$\mathbf{S}_{f1,2,2,3} = [4, -4, 4, 4, 4, -4, 4, 4, 4, -4, 4, 4, -4, 4, -4, -4]^T$$
$$f_{1,2,2,3}(x_1, x_2, x_3, x_4) = x_3 \oplus x_3x_4 \oplus x_1x_2$$

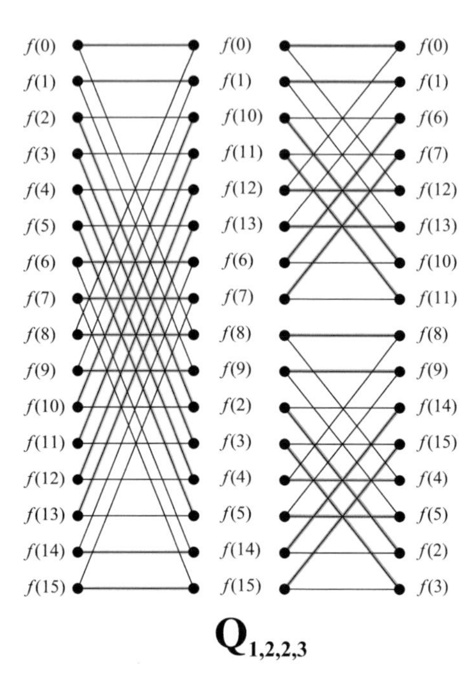

Fig. 8 Flow-graph for computing with $\mathbf{Q}_{1,2,2,3}$

The same function can be obtained by the spectral invariant operation $x_2 \rightarrow x_3 \oplus x_3$ followed by $x_1 \rightarrow x_1 \oplus x_2$. □

8 Algorithm for Constructing Bent Functions

From the proposed approach, an algorithm for constructing bent functions with selected number of variables and specified number of non-zero values in the function vector can be directly formulated.

Determine a library $\mathcal{L}$ of FFT-like permutation matrices for specified values of the number of variables n.

1. Given an arbitrary bent function f with the selected number of non-zero values.
2. Apply an FFT-like permutation matrix to f and write the result in a list.
3. Repeat the Step 2 for different permutation matrices from $\mathcal{L}$.
4. Check if each of the produced new bent functions is already contained in the list. If *yes* skip it and return to Step 2.

Features of the proposed method can be briefly summarized as follows. Referring to spectral invariant operations guaranties that new bent functions are bent and there is no need to check them for bentness, which simplifies the procedure.

Direct implementation of spectral invariant operations requires computing the Walsh spectrum, performing spectral invariant operations over subsets of spectral coefficients, and finally computing the inverse Walsh transform to get the function obtained. Alternative is to deal with functional expressions and perform symbolic computations. By relating permutations determined by spectral invariant operations to factor matrices describing steps of FFT allows their implementation directly in the functional domain avoiding computing the spectrum and the inverse transform. Implementation of the permutation matrices should be simple, since they are derived from fast computing algorithms.

9 Closing Remarks

Bent functions are an interesting mathematical object due to their peculiar properties. At the same time, there are important applications where bent functions are used. The fast Fourier transform (FFT) is also a very important concept, since by ensuring computation efficiency in both space and time, this algorithm made possible application of powerful mathematical method based on the Fourier analysis and various generalizations in scientific and engineering practice. In this chapter, we established a link between these two concepts by pointing out the following:

1. Various bent functions in n variables, a selected degree d, and the specified number of non-zero values w_1 or w_2 can be constructed from a given bent function with the same parameters n, d, and w_i by permutation matrices expressing a structure equal to the structure of factor matrices describing steps in the Cooley-Tukey FFT based on the Good-Thomas factorization of the Walsh transform matrix, which is the FWT for the Hadamard ordering.
2. The basic transform matrix $\mathbf{W}(1)$ is replaced by the basic permutation matrix $\mathbf{P}(1)$.
3. These permutation matrices correspond to the disjoint spectral translation which is a guarantee that their application preserves the flat spectrum, i.e., the bentness.
4. Permutation matrices corresponding to spectral translation in terms of different combinations of variables are obtained through an auxiliary symbolic matrix sharing the same structure as the factor matrices in the FFT, and a suitable replacement of elements of this matrix by 0 and 1 to get the permutation matrices.

In this way, by using different permutation matrices derived from the FFT and their combinations, both applied in various order, all bent functions for a given n, degree d, and the number of non-zero values w_i can be constructed by starting from an arbitrary bent function with the same n, d, and w_i. This follows from the spectral invariant operations related to the used permutation matrices, since no other known transformations preserve the bentness.

References

1. Beauchamp, K.G.: Walsh Functions and Their Applications. Academic Press, New York (1975)
2. Beauchamp, K.G.: Applications of Walsh and Related Functions with an Introduction to Sequency Theory. Academic Press, Bristol (1984)
3. Besslich, Ph.W.: Determination of the irredundant forms of a Boolean function using Walsh-Hadamard analysis and dyadic groups. IEE J. Comput. Dig. Technol. **1**, 143–151 (1978)
4. Besslich, Ph.W.: Efficient computer method for XOR logic design. IEE Proc. E **129**, 15–20 (1982)
5. Besslich, Ph.W.: Spectral processing of switching functions using signal flow transformations. In: Karpovsky, M.G. (ed.) Spectral Techniques and Fault Detection. Academic Press, Orlando (1985)
6. Besslich, Ph.W., Lu, T.: Diskrete Orthogonaltransformationen. Springer, Berlin (1990)
7. Cooley, J.W., Tukey, J.W.: An algorithm for the machine calculation of complex Fourier series. Math. Comput. **19**, 297–301 (1965)
8. Cusick, T.W., Stănică, P.: Cryptographic Boolean Functions and Applications. Academic Press, San Diego (2009)
9. Good, I.J.: The interaction algorithm and practical Fourier analysis. J. R. Stat. Soc. B **20**, 361–372 (1958). Addendum, Vol. 22, 1960, 372–375.
10. Good, I.J.: The relationship between two fast Fourier transforms. IEEE Trans. Comput. **C-20**, 310–317 (1971)
11. Hurst, S.L.: Logical Processing of Digital Signals. Crane Russak and Edward Arnold, London and Basel (1978)
12. Hurst, S.L., Miller, D.M., Muzio, J.C.: Spectral Techniques for Digital Logic. Academic Press (1985)
13. Muller, D.E.: Application of Boolean algebra to switching circuits design and to error detection. IRE Trans. Electron. Comput. **EC-3**, 6–12 (1954)
14. Reed, S.M.: A class of multiple error correcting codes and their decoding scheme. IRE Trans. Inf. Theory **PGIT-4**, 38–49 (1954)
15. Thomas, L.H.: Using a computer to solve problems in physics. In: Application of Digital Computers, Boston, Mass., Ginn (1963)
16. Tokareva, N.: Bent Functions - Results and Applications to Cryptography. Elsevier, Amsterdam (2015)
17. Wu, Ch.-K., Feng, D.: Boolean Functions and Their Applications in Cryptography. Advances in Computer Science and Technology. Springer, Berlin (2016), ISBN 987-3-662-48863-8, ISBN eBook 978-3-662-48865-2
18. Zhegalkin, I.I.: O tekhnyke vychyslenyi predlozhenyi v symbolytscheskoi logykye. Math. Sb. **34**, 9–28 (1927), in Russian
19. Zhegalkin, I.I.: Arifmetizatiya symbolytscheskoi logyky. Math. Sb. **35**, 311–377 (1928), in Russian

Nonlinear Codes for Test Patterns Compression: The Old School Way

Jan Schmidt and Petr Fišer

1 Introduction

When a digital device is tested, one of the problems is to deliver test stimuli economically. Suppose the device is equipped with n scan chains. Then, the n-bit test vectors must be delivered from outside (from a tester), or generated internally.

A substantial help comes from the fact that test stimuli have large redundancy (don't care bits) [1], and can be efficiently compressed. Two major test delivery architectures employ this fact.

The first approach uses a vector stream with the same number of vectors but with minimum redundancy (and therefore with minimum width). The stream is then processed by a combinational circuit often called *combinational expander* or *combinational decompressor*. Generally, there are no additional requirements on the Boolean functions the expander performs, except to deliver the required vectors.

Another approach is to let an FSM generate the required vectors or their superset. For this to work, the set of vectors must have certain properties, e.g., to be a subset of a linear space. Many FSM classes have been used for this purpose, mostly an LSFR, but also Cellular Automata (CA) [10], or Registers with Non-Linear Update (RNLUs) [7]. Such methods are characteristic for Built-In Self-Test (BIST) applications. In all these cases, the properties of the FSM transition function are of concern.

These two approaches are just extremes of a broad spectrum; combined methods such as reseeding [6], bit-flipping [17], bit-fixing [15, 16], Embedded Deterministic Test [13], etc., are numerous. As they have state, they are commonly called *sequential decoders*.

J. Schmidt (✉) · P. Fišer
Czech Technical University in Prague, Prague 6, Czech Republic
e-mail: jan.schmidt@fit.cvut.cz; petr.fiser@fit.cvut.cz

R. Drechsler, D. Große (eds.), *Recent Findings in Boolean Techniques*,
https://doi.org/10.1007/978-3-030-68071-8_6

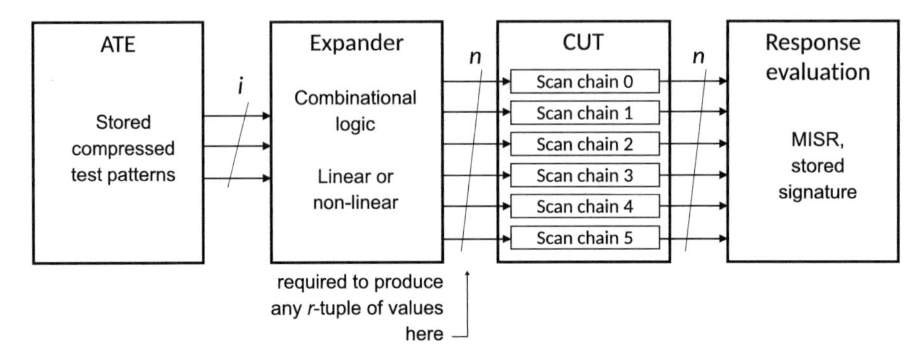

Fig. 1 An example test scenario for three ATE channels and six scan-chains

The compression scheme can be either application dependent or universal for a class of applications with the same number of scan chains and (approximately) the same redundancy in their test sets. The former is undesirable, as the construction of a good compression scheme can be demanding.

Kim and Mitra [9] abstracted the test vector redundancy and required that the expander must be able to set any r-tuple of the produced vector to arbitrary values. Then, the expander is specified only by the number n of scan chains (test vectors width) and the number r. This abstraction has been used by other authors afterward [11, 12].

In this contribution, we will limit ourselves to combinational expanders specified by the above requirements. An example test scenario is shown in Fig. 1. Here compressed test patterns are stored in the tester device (ATE) with i channels. These patterns are then decompressed on-chip to be fed to n ($n > i$) scan chains of the Circuit-under-Test (CUT). Responses to these patterns are then evaluated (typically also on-chip).

Kim and Mitra [9] also brought the idea that introducing redundancy corresponds to encoding a symbol in an error-correcting code. They showed that linear codes, such as BCH codes, are effective for this purpose, and that the r-bit requirement can be translated to a Hamming distance requirement for the dual code. This way, expanders can be easily constructed using existing code tables. They list compressed vector lengths i that can satisfy the r-bit requirement in n scan chains, using an undisclosed linear code for $r = 3$ or BCH codes for $4 \leq r \leq 8$, or Reed-Solomon codes for large n in the case of identified clusters in the test vectors.

An extension to codes other than linear is obvious. It gives much more freedom to choose the expander function, but such freedom also translates to a much bigger search space. Moreover, there is not such a wealth of existing knowledge as in the case of linear codes. And, last but not least, nonlinear codes that are efficient for error corrections are not guaranteed to be efficient for test vector expansion.

Many efforts come from the BIST domain. Dutta and Touba [4] limit the search space by considering only a limited class of circuits. Novák [11] extends linear codes by nonlinear expander outputs. The stochastic search over the complete search space in [12] brought functions that are remarkably efficient. The authors generate a truth table randomly under certain stochastic requirements and then check the r-bit requirement. Functions for larger r are then composed.

These researches seem to state that "with i tester channels and the r-bit requirement, my code can accommodate up to n scan chains." We believe that the question in design time is rather "with n scan chains and the r-bit requirement, what is the minimum number i of tester channels?". While the optimization task is the same, the difference must be regarded when comparing.

Definition 1 (Expander Function) Given integers i, n, r, where $r < n$ and $i < n$, an *expander function* is a function $f_{i,n,r} : \{0, 1\}^i \rightarrow \{0, 1\}^n$ such that for all ordered r-tuples P of positions from $[0, n-1]$ and for all valuations $V \in \{0, 1\}^r$ of these positions, there exists a vector $\mathbf{x} \in \{0, 1\}^i$ such that the vector $f_{i,n,r}(\mathbf{x})$ has the values V at positions P.

Definition 2 (Expander Minimization) Given integers n, r, $r < n$, find the smallest integer i such that there exists an expander function $f_{i,n,r}$.

In the proposed approach, we formulate all requirements to the expander function first. Any function that satisfies the requirements is therefore a correct solution. Then, we use synthesis tools to get an optimized implementation of the expander.

The chapter is organized as follows. In Sect. 2 we formulate the problem as a clique cover problem, followed by multi-valued (MV) variable encoding and logic synthesis. We analyze instance properties in Sect. 2.2 and outline a simple heuristic in Sect. 2.3. The concrete methods used and their results are described in Sect. 3. We outline results we hope for in Sect. 4.

2 Proposed Approach

We specify all outputs the expander must produce as the output part of its (incompletely specified) function. Then, we are free to construct any input part to optimize the circuit without affecting the correctness of function. Last, we synthesize the circuit.

2.1 Expander Outputs as a Clique Cover Problem

We construct requirements (or constraints) on the expander first. The r-bit requirement tells us that, for every r-tuple of the n expander outputs, all 2^r combinations of binary values must be present for at least one input value. Such constraint can be expressed as 2^r cubes of dimension $n - r$, called *requirement cubes*. The collection of all such cubes completely specifies the output of the expander as the set of *output cubes* of its expander function. Formally,

Definition 3 (Requirements Cube) Given integers n, r, where $r < n$, an ordered r-tuple P of positions from $[0, n-1]$, valuations $V \in \{0, 1\}^r$ of these positions, a *requirement cube* $\rho_{P,V}$ for P and V is a subset $\rho_{P,V} \subset \{0, 1, -\}^n$, obtained by setting values from V in dimensions given by P, and don't-cares (DC, '-') otherwise. It is therefore a cube of dimension $n - r$.

Definition 4 (Requirements Set) Given integers n, r, where $r < n$, we call the set R^n_r of all requirement cubes $\rho_{P,V}$ for all P r-tuple and all valuations $V \in \{0, 1\}^r$ a *requirement set* for n and r.

During the operation of the expander, we do not need to distinguish between all requirement cubes. For every valuation of every r-tuple, there must be an input to the expander that produces those values at the output. Therefore, cubes that intersect can be replaced by their intersection. The reduced number of distinguished outputs saves the resulting expander width.

Understanding the intersection as compatibility, we can construct a compatibility graph and then treat it as a covering problem. All compatible cubes form a clique. As the intersection of two cubes is a cube, each clique also corresponds to a cube.

Definition 5 (Clique Characterization) Let R^n_r be a requirements set, and P a set of mutually compatible requirement cubes, a clique. Then we say that a cube c constructed as the intersection of all cubes in P *denotes* the clique P.

Any clique cover of the graph is a valid set of all output vectors of the expander function. The input width i is the logarithm of the number of output vectors and does not depend on further optimizations. Therefore, we seek a minimum cover.

Figures 2 and 3 show a rather small example. There are $N_R = 24$ requirement cubes. Each of the cubes is compatible with 12 other cubes. There are 16 distinct minimum covers of size 5, hence their cliques are identified with $i = 3$ bits. No clique in the covers can be replaced by a smaller clique. The solutions form one class of equivalence, as defined below.

```
00--  0--0  -0-0
01--  0--1  -0-1
10--  1--0  -1-0
11--  1--1  -1-1

0-0-  -00-  --00
0-1-  -01-  --01
1-0-  -10-  --10
1-1-  -11-  --11
```
a

```
0000  1000
0001  1001
0010  1010
0011  1011
0100  1100
0101  1101
0110  1110
0111  1111
```
b

```
0001
0010
0100
1000
1111
```
c

```
v1| 0001
v2| 0010
v3| 0100
v4| 1000
v5| 1111
```
d

```
000| 0001
001| 0010
010| 0100
011| 1000
100| 1111
```
e

Fig. 2 Description of the $n = 4, r = 2$ example: (**a**) the set R^2_4 of requirement cubes, (**b**) cubes denoting all cliques, (**c**) cubes denoting a minimum clique cover, (**d**) a possible multi-valued specification of the expander function (undefined at the points not in the table), (**e**) possible encoding of the MV specification

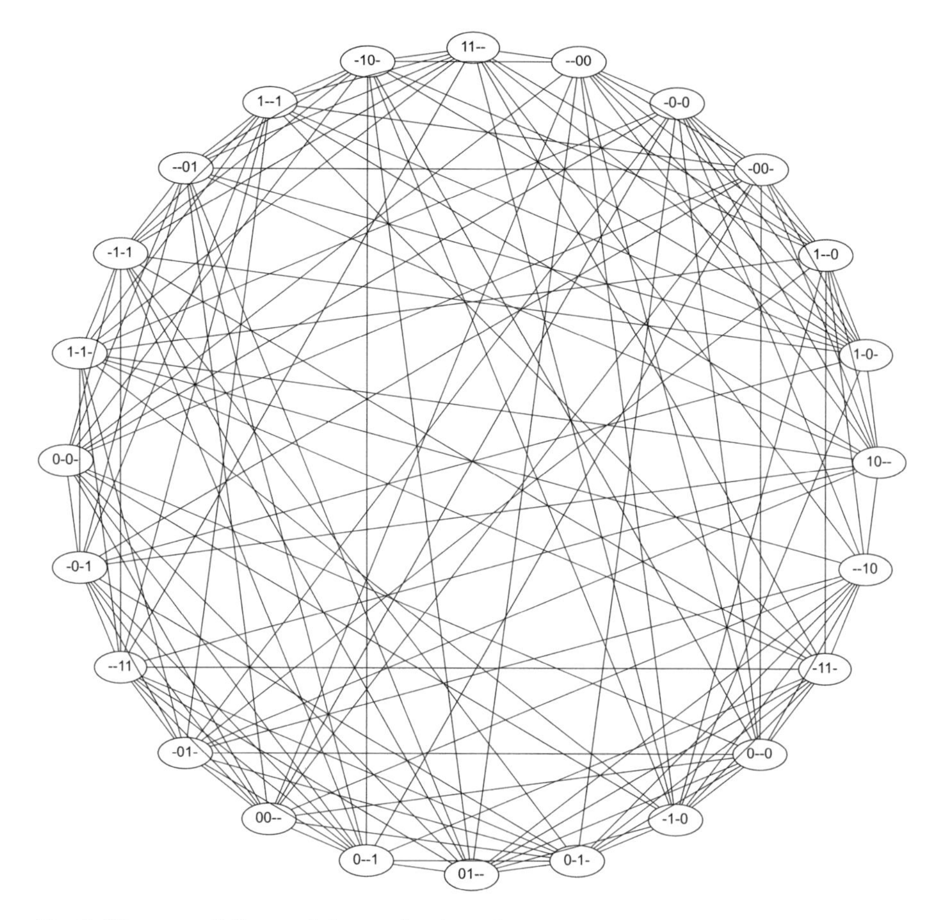

Fig. 3 The compatibility graph for $n = 4$ and $r = 2$

To obtain the Boolean function of the expander, it is sufficient to number the cubes and encode the numbers in any way. Then, the encoded numbers will form the input part of incomplete function specification, and the cubes the output part. Technically, this is a PLA-format specification, as defined by two-level minimizers [14].

2.2 *Compatibility Graph Properties*

By construction, the compatibility graph has

$$N_R = \binom{n}{r}.2^r \tag{1}$$

nodes. In the example in Figs. 2 and 3, $N_R = \binom{4}{2}.2^2 = 24$.

Let R_r^n be a requirements set and let $\rho \in R_r^n$ be a requirement cube. Now let us count other compatible requirement cubes. By construction, ρ contains r care bits. To be compatible, a cube must contain either the same values or don't cares at those places. Let us choose j places of ρ. We have $\binom{n}{r}$ possibilities. As the compatible cube must have r care places, there are $\binom{n-r}{r-j}$ possibilities for the positions of the remaining $r - j$ care bits. As each of them may be 0 or 1, we finally get the number N_{COMPAT} of cubes compatible with any given cube as

$$N_{COMPAT} = \sum_{j=0}^{r-1} \binom{r}{j} \binom{n-r}{r-j} 2^{r-j} \tag{2}$$

In the example in Figs. 2 and 3, $N_{COMPAT} = \binom{2}{0}\binom{2}{2}.2^2 + \binom{2}{1}\binom{2}{1}.2 = 12$.

Theorem 1 *Let c be any cube of dimension $k \leq n - r$. Then c denotes a clique of size $S_C(k) = \binom{n-k}{r}$. The number of such cliques is $N_C(k) = \binom{n}{n-k}2^{n-k}$.*

Proof By construction, any cube c of dimension $k \leq n-r$ denotes a clique, as there are requirement cubes compatible with c. The number of specified variables in c is $n - k$. there are $S_C(k) = \binom{n-k}{r}$ r-tuples having the same specified variables as c. There are $\binom{n}{n-k}$ combinations of variables in c, each of which can have two values, hence $N_C(k) = \binom{n}{n-k}2^{n-k}$. □

Let us note that especially $N_C(0) = 2^n$. In the example in Figs. 2 and 3, $N_C(0) = 2^4 = 16$.

Theorem 2 *Let R_r^n be a requirements set and let C be a clique cover of R. Then, any clique denoted by a cube of dimension k, $n - r \geq k > 0$ can be replaced by a clique denoted by a cube of dimension 0, without changing the size of C.*

The proof follows from Theorem 1. Reducing the dimension of a cube means enlarging the corresponding clique, which is permissible because the problem is *unate covering* in the sense of Brayton et al. [3].

From Theorem 2, it follows that among minimum size covers, at least one is composed of cliques denoted by dimension-0 cubes. It may happen that we need to go the other way round: Given a clique cover denoted by a set of dimension-0 cubes, can the dimensions of some cubes be enlarged?

Definition 6 (Clique-Reducibility) Let R_r^n be a requirements set and let C be a clique cover of R. If any clique in the cover can be replaced by a clique denoted by a cube of a larger dimension, such cover is called *clique-reducible*, otherwise it is *clique-irreducible*.

The solution presented in Fig. 2 is an example of a clique-irreducible cover.

Theorem 3 *Let R_r^n be a requirements set for any n and $r \leq n$. Let $T : \{0, 1, -\}^n \rightarrow \{0, 1, -\}^n$ be a function with the following properties:*

- *T is a bijection on R.*

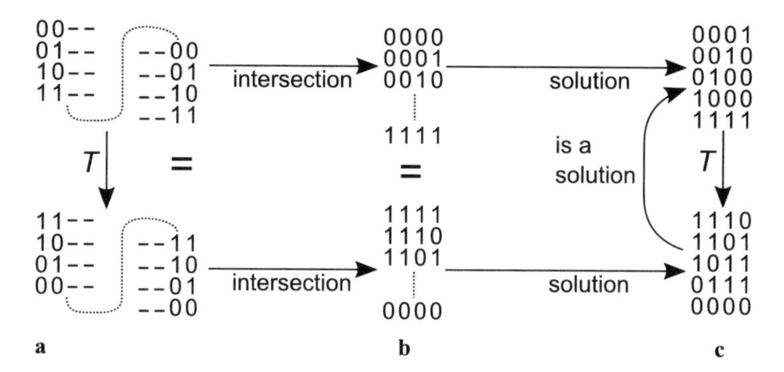

Fig. 4 Example transformation of the $n = 4, r = 2$ example, with all values' negation as T: (**a**) the set R_4^2 of requirement cubes, (**b**) cubes denoting all cliques, (**c**) cubes denoting a minimum clique cover

- *T preserves intersection:* $\forall c_1, c_2 \in \{0, 1, -\}^n, T(c_1) \cap T(c_2) = T(c_1 \cap c_2)$.

Then, if the set $C = \{c_1, c_2, \ldots, c_m\}$ *denotes a clique cover, then the set* $T(C) = \{T(c_1), T(c_2), \ldots, T(c_m)\}$ *also denotes a set cover.*

Proof Because T is a bijection (permutation) on R, $T(R_r^n) = R_r^n$. Because it preserves intersection, also the clique sets are the same. If some cubes c_1 and c_2 intersect (are compatible), then so do $T(c_1)$ and $T(c_2)$. By construction, T is also a bijection on $\{0, 1, -\}^n$. Therefore, $T(C)$ is also a clique cover on R_r^n. □

Figure 4 illustrates the role of T.

Theorem 4 *A function* $T : \{0, 1, -\}^n \to \{0, 1, -\}^n$*, which satisfies the antecedent of Theorem 3, also preserves clique-reducibility.*

Definition 7 (Cube Position Inversion) Let c be a cube in $\{0, 1, -\}^n$, and k an integer, $1 \leq k \leq n$. Then $inv(c, k)$ is a cube obtained from c by inverting a care value at position k, or c if there is a don't care value at position k.

Definition 8 (Cover Column Inversion) Let C be a set of cubes in $\{0, 1, -\}^n$, and k an integer, $1 \leq k \leq n$. Then $inv(C, k)$ is a set of cubes $\{\forall c \in C : inv(c)\}$.

Definition 9 (Cube Permutation) Let c be a cube in $\{0, 1, -\}^n$ and let P be a permutation of the sequence $1 \ldots n$. Then $perm(c, P)$ is a cube obtained from c by permuting all positions of c according to P.

Definition 10 (Cover Column Permutation) Let C be a set of cubes in $\{0, 1, -\}^n$ and let P be a permutation of the sequence 1 . . . n. Then $perm(C, P)$ is a set of cubes $\{\forall c \in C : perm(c, P)\}$.

Theorem 5 *The transformation* $inv(C, k)$ *satisfies the requirements of Theorem 3 for* $1 \leq k \leq n$.

Proof If the literal at position k in a cube c is a don't care, c transforms to itself. If it is a care value, then there is exactly one cube in R_r^n with inverted value and those two cubes map to each other. Therefore, $inv(C, k)$ is a bijection. If the cubes c_1 and c_2 are compatible, then so are $inv(c_1, k)$ and $inv(c_2, k)$. Because $inv(c, k)$ is an inverse to itself, an equivalence takes place and $inv(C, k)$ preserves intersection. □

Theorem 6 *The transformation $perm(C, P)$ satisfies the antecedent of Theorem 3, for all P.*

Proof The antecedent of Theorem 3 does not depend on position ordering, as long as it is consistent over R_r^n and therefore over all cliques. □

Theorem 7 $\forall C, 1 \leq k \leq n : inv(perm(C, P)) = perm(inv(C, k), P)$.

Therefore,

Theorem 8 *Let R be a set of all requirement cubes for given n and r. Then, for any set of cubes C, $1 \leq k \leq n$, $\{C, inv(C), \forall P : perm(C, P), \forall P : perm(inv(C, k), P)\}$ is an equivalence class with respect to covering R_r^n.*

Equation (2) and Theorem 8 state that the problem is highly regular and symmetric, which gives some hope to find either the minimum clique cover size N_{OPT} analytically in the future, or to use Theorem 8 to prune a search efficiently.

2.3 Techniques for Minimum Clique Cover

Very small instances can be solved by brute force, which can give some insight into the problem. Because we know that the covers are not large for small instances, we constructed Algorithm 1, which tries to construct a cover of a given size s. Furthermore, it uses only cubes of dimension 0 (that is, completely defined) for the cover.

As (1) shows, compatibility graphs tend to be large even for small n and r. Therefore, we sought a way to get the clique cover without *storing* requirement cubes explicitly, in an on-line fashion. This, of course, does not guarantee optimality, as it is a sort of greedy technique. The procedure can be outlined as Algorithm 2.

Notice that this is still a kind of meta-algorithm: the order of tuple selection in Line 2 and values selection in Line 3 is not defined. Also, the algorithm is satisfied with the first candidate clique that covers a given cube (Line 7).

2.4 Expander Input as an MV-Encoding Problem

With a given cover C, the construction of the expander function still has a great degree of freedom. Only $|C|$ values out of the 2^i are used. By the construction of i,

Algorithm 1 Cube generation and exact clique cover

Input: n, r, s ▷ s is the tried cover size
Output: A set $\{C\}$ of all clique covers C
1: Let R be an empty set of requirement cubes.
2: **for all** ordered r-tuples P_R from $1 \ldots n$ **do**
3: **for all** $V \in \{0, 1\}^n$ **do**
4: construct a cube q having values from V at places P_R and don't cares otherwise.
5: insert q into R.
6: **end for**
7: **end for**
8: Let $\{C\}$ be an empty set of covers.
9: **for all** ordered s-tuples P_C from $1 \ldots 2^n$ **do**
10: let C be an empty cover.
11: **for all** members p of P_C **do**
12: construct a cube c from the binary representation of p
13: insert c into C
14: **end for**
15: **if** C covers R **then**
16: insert C into $\{C\}$.
17: **end if**
18: **end for**

Algorithm 2 Cube generation and greedy clique cover

Input: n, r
Output: A clique cover C.
1: Let C be empty.
2: **for all** ordered r-tuples P from $1 \ldots n$ **do** ▷ in some order
3: **for all** $V \in \{0, 1\}^n$ **do** ▷ in some order
4: construct a cube q having values from V at places P and don't cares otherwise
5: **for all** cubes $c \in C$ **do**
6: **if** c is compatible with q **then**
7: replace c with $c \cap q$
8: **break**
9: **end if**
10: **end for**
11: **if** q still uncovered **then**
12: insert q into C
13: **end if**
14: **end for**
15: **end for**

$2^{i-1} < |C| \leq 2^i$. It means that up to half of the values are unused, and therefore belong to the DC set of the expander function.

More importantly, once we have the output part of the two-level specification of the expander function, we are free to choose a distinct input pattern for each of the output cubes. We can treat the expander inputs as a multi-valued (MV), symbolic variable. Then the problem is how to encode it to produce a minimum circuit. At a first glance, this is similar to the opcode encoding or FSM state encoding problem [14]. The difference, especially from the state encoding problem, is that here the variable is an external input variable. Available encoding algorithms cannot benefit from the degree of freedom. Nevertheless, the constraints from such encoders can be used in a specialized algorithm.

2.5 Method Summary

Given the encoder specification, the expander can be synthesized. Starting the synthesis by a two-level description, which does not suggest circuit structure, is unusual in contemporary practice (cf. [5]). In this situation, the classical minimization-decomposition approach, e.g., using BDS [18], seems worth consideration. However, any contemporary logic optimization tool accepting such a description at its input can be used, e.g., ABC [2].

Our method can be outlined as Algorithm 3. An example description obtained in Steps 10 and 11 is in Fig. 5. Note that Fig. 5a describes an incompletely specified function; the on-set and off-set are specified explicitly, while minterms that are not listed are assigned don't cares. Figure 5b shows a minimized, completely specified function description. Here the "$\sim$" symbols in the output have no value meaning; only "1"s indicate the on-set. The complement to this on-set is the off-set. These two descriptions indeed correspond to the Espresso PLA formats *fr* and *fd*, respectively [14].

Algorithm 3 Method overview

1: For every value of every r-tuple in an n bit vector, construct a *requirement cube* ρ stating that those values are output, and nothing else.
2: Construct a *compatibility graph* $G(R, E)$, such that R is the set of all requirement cubes and there is an edge between requirements cubes ρ_1 and ρ_2 iff they intersect.
3: Solve Minimum Clique Cover on G. Let the number of cliques be N_{OPT}.
4: The necessary input width is $i = \lceil \log_2 N_{OPT} \rceil$.
5: Collect all clique cubes as the output part of the two-level description.
6: **if** MV optimization is available: **then**
7: Choose N_{OPT} distinct symbolic values for the left-hand side, providing an MV description.
8: Perform MV minimization and encoding on the MV description.
9: **else**
10: Choose N_{OPT} distinct binary combinations for the input part, providing a two-level description.
11: Perform minimization of the two-level description.
12: **end if**
13: Synthesize the resulting two-level description by any method, possibly technology-dependent.

```
0000 | 0000000000
0001 | 0000111110
0010 | 1111000001
0011 | 1111111111
0100 | 0110000111
0101 | 1001111000
0110 | 0101010011
0111 | 1010101100
1000 | -0--0-1--1
1001 | -1--1-0--0
```

a

```
1--0 | ~~~~~~1~~1
1--1 | ~1~~1~~~~~
00-1 | ~~~~11111~
-100 | ~11~~~~111
-101 | 1~~1111~~~
-110 | ~1~1~1~~11
-01- | 1111~~~~~1
--11 | 1~1~1~11~~
```

b

Fig. 5 Two-level description of an $n = 10$, $r = 2$ expander, (**a**) as generated by Algorithm 2 (undefined at the points not in the table), (**b**) after minimization (e.g., by Espresso)

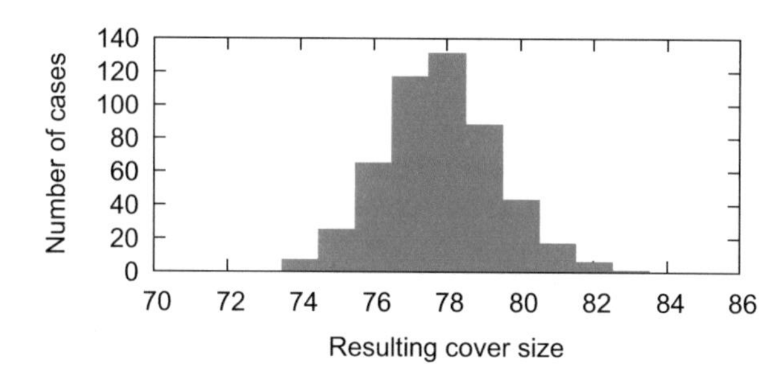

Fig. 6 N_{SUB} frequencies in 500 runs of Algorithm 2 for $n = 20$ and $r = 4$

3 Results

With the techniques described above, we solved instances up to $n = 32$ and $r = 6$. Besides the comparison of the resulting widths, we also investigated the importance of good MV encoding, synthesis approaches, and expander circuits properties.

3.1 Implementation

Algorithms 2 and 1 have been implemented as a sequential C++ program and run on an office machine (Intel i7, 8 cores at 3GHz, SUSE Linux). The ordering in Line 2 of Algorithm 2 has been implemented as ordered random sampling [8, p. 166]. The ordering in Line 3 is systematic. A variant of the algorithm, which found a cube c maximally intersecting p (Line 7) has been tested with no improvement.

To illustrate the influence of randomization, Algorithm 2 has been run 500 times for $n = 20$ and $r = 4$. The resulting histogram is in Fig. 6. Random selection does have an influence on the obtained suboptimum cover size N_{SUB}; however, it is unlikely that the differences cause a difference in i.

Table 1 Exact and heuristic solutions of small instances

		Exact				Heuristic		
n	r	i	N_{OPT}	Number of solutions	Number of classes	i	N_{SUB}	N_{SUB}/N_{OPT}
4	2	3	5	16	1	3	6	1.20
4	3	3	8	2	1	4	12	1.50
5	2	3	6	2896	7	3	8	1.33
5	3	4	10	16	1	4	15	1.50
6	2	3	7	2,080,192	4	3	8	1.14
7	2	3	8	2,845,462	–	3	8	1.00

3.2 Resulting Codes

Table 1 compares Algorithms 1 and 2. We can see that the heuristic operates within 150% of the optimum cover size. The number of solutions and classes differ wildly, indicating that the characters of the instances also differ. It seems that the gap between greedy and exact solutions closes with larger instances.

Table 2 summarizes instances and results obtained from Algorithm 2. Although instance parameters show that the element counts grow rapidly, clique cover sizes remain relatively small.

Table 3 contains expander widths i reported by various authors in comparison to results of Algorithm 2. The table is limited to comparable values of n and r, so that results in the practical range of n in the hundreds are not included. The presented algorithms, due to combinatorial explosion, will never be able to work in that range.

The results in [12] seem to form two groups: one with n above 40, and the other group below. The proposed algorithm is able to match the second group. It also outperforms linear BCH codes and optimum linear codes for larger n.

3.3 Is Optimum MV Encoding Important?

We have found no way to use MV optimizers to design compressed test encoding, that is, the input part of the two-level description. For the optimizers, the symbolic input part is an external, given information. To estimate how important the encoding is, we arranged the following experiment.

For every code obtained by Algorithm 2, 500 input parts were generated randomly. Each two-level description was optimized by Espresso [14]. The resulting minimized descriptions were characterized by the total number of literals. Statistical characterization of the result is in Table 4.

The expected result is that the size of the encoder depends on input encoding. This happens at least in some cases—in the case of $n = 10$ and $r = 2$, the difference between the worst and the best result spans 66% of the median. The surprising fact is

Table 2 Instance properties, resulting widths i and achieved cover sizes N_{SUB} from Algorithm 2

n	r	N_R	S_C	i	N_{SUB}	n	r	N_R	S_C	i	N_{SUB}
10	2	180	45	4	10	28	2	1512	378	4	14
10	3	960	120	5	21	28	3	26,208	3276	6	35
10	4	3360	210	6	52	28	4	327,600	20,475	7	92
10	5	8064	252	7	118	28	5	3,144,960	98,280	8	231
10	6	13,440	210	8	215	28	6	24,111,360	376,740	10	569
16	2	480	120	4	12	30	2	1740	435	4	14
16	3	4480	560	5	27	30	3	32,480	4060	6	37
16	4	29,120	1820	7	69	30	4	438,480	27,405	7	97
16	5	139,776	4368	8	165	30	5	4,560,192	142,506	8	244
16	6	512,512	8008	9	365	30	6	38,001,600	593,775	10	596
20	2	760	190	4	12	32	2	1984	496	4	14
20	3	9120	1140	5	30	32	3	39,680	4960	6	37
20	4	77,520	4845	7	74	32	4	575,360	35,960	7	98
20	5	496,128	15,504	8	194	32	5	6,444,032	201,376	8	248
20	6	2,480,640	38,760	9	443						
24	2	1104	276	4	14						
24	3	16,192	2024	6	34						
24	4	170,016	10,626	7	84						
24	5	1,360,128	42,504	8	211						
24	6	8,614,144	134,596	10	513						

Table 3 Expander widths comparison

		Width (i) for code							Width (i) for code				
n	r	Linear [9, 12]	BCH [9]	NBC1 [12]	NBC2 [12]	Proposed	n	r	Linear [9, 12]	BCH [9]	NBC1 [12]	NBC2 [12]	Proposed
8	3	4				5	12	5	8				8
12	3	4				5	13	5				7	8
16	3	5				5	18	5			8		8
32	3	6				6	24	5	10			8	8
8	4	6				5	11	6	9				8
10	4	7				6	13	6			9		9
13	4	8				6	16	6			9		9
14	4	6		6		6	17	6	10				9
18	4	6			6	7	32	4		12			7
9	5	7				7	32	5		18			8
10	5			7		7	32	6		18			15

Table 4 Number of literals after minimization, random encodings. m is the median of the distribution, and Δ is the difference between max and min

n	r	min	m	max	σ/m	Δ/m	n	r	min	m	max	σ/m	Δ/m
10	2	35	57	73	0.07	0.667	28	2	134	182	216	0.047	0.451
10	3	114	153	178	0.052	0.418	28	3	508	564	606	0.013	0.174
10	4	339	412	501	0.041	0.393	28	4	1637	1738	1773	0.009	0.078
10	5	946	1063	1272	0.026	0.307	28	5	4747	4851	4892	0.003	0.03
10	6	1870	2040	2408	0.017	0.264	28	6	12,169	12,293	12,390	0.002	0.018
16	2	68	100	119	0.05	0.51	30	2	132	193	228	0.028	0.497
16	3	247	298	325	0.023	0.262	30	3	579	662	697	0.021	0.178
16	4	723	845	889	0.017	0.196	30	4	1834	1951	2002	0.009	0.086
16	5	1987	2251	2323	0.009	0.149	30	5	5244	5398	5465	0.004	0.041
16	6	5333	5474	5594	0.005	0.048	30	6	13,331	13,474	13,574	0.002	0.018
20	3	329	385	413	0.019	0.218	32	3	620	704	738	0.015	0.168
20	4	1018	1090	1112	0.008	0.086	32	4	1988	2090	2128	0.007	0.067
20	5	3043	3176	3242	0.006	0.063	32	5	5666	5787	5850	0.003	0.032
20	6	7748	7882	7970	0.003	0.028							
24	3	428	497	529	0.025	0.203							
24	4	1314	1412	1452	0.009	0.098							
24	5	3852	3954	4002	0.004	0.038							
24	6	9791	9929	10,004	0.002	0.021							

that this influence rapidly diminishes with increasing r. There is a strong correlation of -0.9. No other such correlation has been found. It is true that the cover does not use all 2^i values, and that the number of unused values differs from instance to instance. Yet the spread does not correlate with that at all (less than 0.1).

3.4 Expander Synthesis

The starting point of the synthesis is a two-level description, which is unusual. We therefore chose two alternative approaches to synthesis. The first one is classical; minimization with Espresso [14] followed by decomposition with BDS [18]. The other approach uses only ABC [2] with a rather high effort. The script in Algorithm 4 is iterative as recommended by the authors of ABC. Nevertheless, we tried to combine both approaches and use ABC with the same script as a post-optimizer for Espresso/BDS. The output of all three approaches is a network of 2-input gates of arbitrary type.

The comparison of the two synthesis approaches shows that although the decomposition approach performance is acceptable, iterated synthesis can provide yet better results. It does not suffer from the absence of structure, which means that it is able to discover the circuit structure independently. ABC as a post-optimizer

Algorithm 4 Expander synthesis using ABC

Input: in.pla: a two-level description of the expander
Output: out.blif: Optimized BLIF description of the expander

```
1: read_pla in.pla; &get -n                          ▷ convert into the GIA structure
2: for 20 times do
3:     &st; &synch2; &if -m -a -K 2; &mfs -W 10;
4:     &st; &dch; &if -m -a -K 2; &mfs -W 10
5: end for
6: &put; write_blif out.blif
```

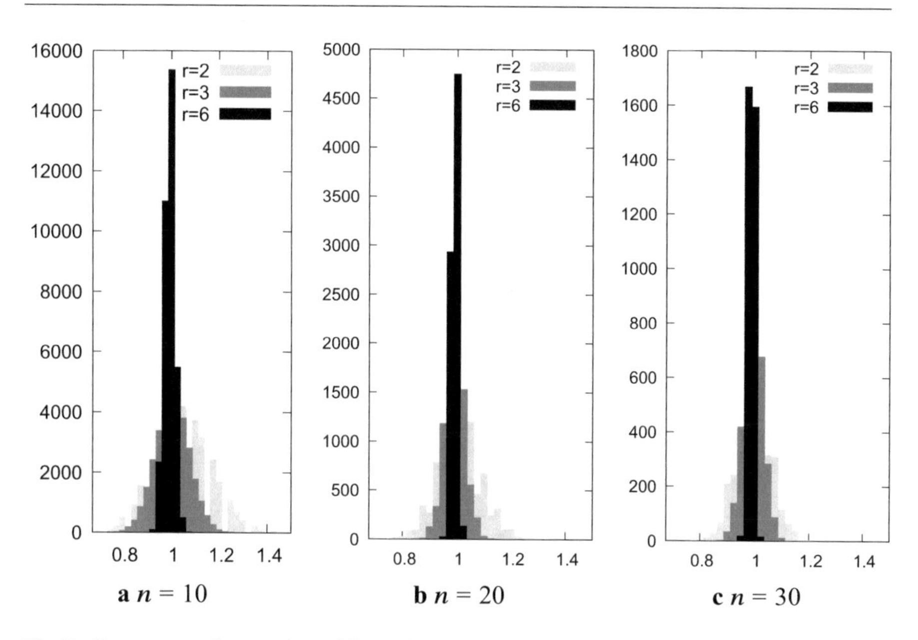

Fig. 7 Gate counts frequencies with random encoding, normalized to median. (**a**) $n = 10$. (**b**) $n = 20$. (**c**) $n = 30$

improves the results of the classical approach considerably, achieving the best results from the three alternatives.

The synthesis results obtained with random encoding corroborate Sect. 3.3 in the sense that the influence of encoding decreases with growing r. As Fig. 7 illustrates, this holds also for increasing n to some degree.

The resulting gate counts in Table 5 indicate that, with current methods, the improved expander width is paid for by expander size. To illustrate, a small experiment for $n = 15$ and $r = 4$ is presented in Table 6. A linear expander using the BCH(4,2) code has been synthesized with Espresso and BDS. A nonlinear expander has been generated using Algorithm 2 and binary input encoding. The synthesis flow was the same. Although the BCH(4,2) code is not optimal, the expander is considerably smaller.

Table 5 Expander implementations, gate counts with random encoding

		ABC			Espresso, BDS			Espresso, BDS, ABC		
n	r	min	median	max	min	median	max	min	median	max
10	2	22	33	49	16	40	69	15	28	48
10	3	69	94	126	74	111	145	59	84	115
10	4	188	228	276	212	260	307	167	207	244
10	5	441	492	552	488	550	610	403	448	495
10	6	889	972	1046	978	1071	1171	792	876	952
16	2	36	50	73	35	69	101	30	48	70
16	3	128	159	194	149	189	232	119	148	177
16	4	498	588	662	643	735	823	497	576	655
16	5	1214	1309	1403	1472	1568	1690	1180	1267	1361
16	6	2625	2770	2907	3057	3201	3374	2471	2592	2726
20	2	39	58	82	45	84	122	36	58	83
20	3	168	198	239	194	234	276	151	183	213
20	4	650	738	817	823	924	1014	654	726	808
20	5	1612	1736	1830	1883	1989	2125	1533	1614	1708
20	6	3572	3700	3854	3888	4044	4226	3168	3276	3419
24	2	52	72	101	56	102	138	47	72	96
24	3	290	355	431	401	492	572	305	360	428
24	4	816	917	1010	1041	1127	1223	815	894	976
24	5	1984	2104	2220	2277	2376	2506	1841	1939	2028
24	6	6427	6645	6848	8065	8360	8702	6335	6570	6866
28	2	58	88	113	77	123	158	59	88	115
28	3	340	408	493	458	567	667	334	414	490
28	4	1001	1092	1195	1234	1323	1420	971	1062	1136
28	5	2387	2512	2615	2659	2772	2904	2154	2256	2372
28	6	7837	8109	8335	9623	9900	10,206	7587	7848	8103
30	2	64	92	122	82	132	168	64	92	120
30	3	386	462	541	530	623	717	366	462	549
30	4	1098	1197	1291	1326	1421	1522	1066	1140	1220
30	5	2604	2718	2852	2842	2970	3127	2310	2408	2517
30	6	8602	8848	9074	10,404	10,700	11,003	8136	8480	8729
32	2	65	94	122	88	138	173	61	96	123
32	3	399	486	574	538	656	748	401	486	559
32	4	1172	1274	1368	1412	1520	1625	1136	1212	1306
32	5	2811	2912	3028	3029	3150	3280	2457	2552	2668

Table 6 A comparison of a linear and a nonlinear expander for $n = 15$ and $r = 4$

	BCH (4,2)	Proposed
Width i	8	7
Literals	108	807
Gates	27	329

4 Future Directions

The main hope of this work is to gain knowledge about the problem. As (2) and Theorem 8 indicate high degree of symmetry in all instances, we hope that the problem is not so difficult as the numbers in Table 2 indicate. Any N_{SUB} gives a lower bound for encoder width. To find N_{OPT} analytically would enable us to judge optimality of any code. Another potential benefit of the high symmetry could result in better construction algorithm.

5 Conclusions

The problem to design a nonlinear code for a test vectors expander, with the requirement of all r-tuples possible in the output, has been formulated as a clique cover problem. The instances of the problem are large but have a high degree of symmetry, which seems to offer the possibility of analytical solution or better heuristic construction. To benefit from the degrees of freedom in the problem, assigning expander inputs to the produced vectors has been identified as a multi-valued (MV) variable encoding problem.

Experimental evaluation shows that good MV encoding is important for small r. For small instances up to $n = 6$ and $r = 2$, the sets of all minimum size covers were obtained using brute force. Instances up to $n = 32$ and $r = 6$ were solved heuristically, with the resulting expander widths i mostly equal to, and sometimes better than, existing solutions. For the synthesis of the expanders, both the classical minimization-decomposition and resynthesis approaches can be used. The produced circuits were larger than corresponding linear expanders.

Acknowledgements Computational resources were supplied by the project "e-Infrastruktura CZ" (e-INFRA LM2018140) provided within the program Projects of Large Research, Development and Innovations Infrastructures. The authors acknowledge the support of the OP VVV MEYS funded project CZ.02.1.01/0.0/0.0/16_019/0000765 "Research Center for Informatics."

References

1. Agrawal, V.D., Kime, C.R., Saluja, K.K.: A tutorial on built-in self-test. 2. Applications. IEEE Des. Test Comput. **10**(2), 69–77 (1993)
2. Berkeley Logic Synthesis and Verification Group. ABC: A System for Sequential Synthesis and Verification. Software System. UC, Berkeley (2000). http://www.eecs.berkeley.edu/~alanmi/abc/
3. Brayton, R.K., et al.: Logic Minimization Algorithms for VLSI Synthesis. Kluwer Academic Publishers, Boston (1984), p. 192
4. Dutta, A., Touba, N.A.: Using limited dependence sequential expansion for decompressing test vectors. In: 2006 IEEE International Test Conference (2006), pp. 1–9
5. Fišer, P., Schmidt, J.: A difficult example or a badly represented one? In: Proceedings of 10th International Workshop on Boolean Problems. Technische Universität Bergakademie, Freiberg (2012), pp. 115–122. ISBN: 978-3-86012-438-3
6. Hellebrand, S., et al.: Built-in test for circuits with scan based on reseeding of multiple-polynomial linear feedback shift registers. IEEE Trans. Comput. **44**(2), 223–233 (1995)
7. Li, N., Dubrova, E.: Area-efficient high-coverage LBIST. Microprocessors Microsyst. **38**(5), 368–374 (2014). ISSN: 0141-9331. https://doi.org/10.1016/j.micpro.2014.05.002. http://www.sciencedirect.com/science/article/pii/S0141933114000738
8. McGeoch, C.C.: A Guide to Experimental Algorithmics. Cambridge University Press, Cambridge (2012), p. 261
9. Mitra, S., Kim, K.S.: XPAND: an efficient test stimulus compression technique. IEEE Trans. Comput. **55**(2), 163–173 (2006)
10. Novák, O.: Pseudorandom, weighted random and pseudoexhaustive test patterns generated in universal cellular automata. In: Lecture Notes in Computer Science 1667 – Dependable Computing – EDCC-3. Springer, Berlin (1999), pp. 303–320
11. Novák, O.: Extended binary nonlinear codes and their application in testing and compression. In: 2017 22nd IEEE European Test Symposium (ETS) (2017), pp. 1–2
12. Novák, O., Rozkovec, M., Plíva, J.: Decompressors using non-linear codes. Microprocessors Microsyst. **76**, 103076 (2020). ISSN: 0141-9331. https://doi.org/10.1016/j.micpro.2020.103076. http://www.sciencedirectcom/science/article/pii/S0141933119306441
13. Rajski, J., et al.: Embedded deterministic test. IEEE Trans. Comput. Aided Des. Integr. Circ. Syst. **23**(5), 776–792 (2004)
14. Rudell, R.L.: MultipleValued logic minimization for PLA synthesis. Tech. rep. M86/65. University of California in Berkeley ERL, June 1986
15. Touba, N.A., McCluskey, E.J.: Synthesis of mapping logic for generating transformed pseudo-random patterns for BIST. In: Proceedings of 1995 IEEE International Test Conference (ITC) (1995), pp. 674–682
16. Touba, N.A., McCluskey, E.J.: Bit-fixing in pseudorandom sequences for scan BIST. IEEE Trans. Comput. Aided Des. Integr. Circ. Syst. **20**(4), 545–555 (2001)
17. Wunderlich, H.-J., Kiefer, G.: Bit-flipping BIST. In: Proceedings of International Conference on Computer Aided Design (1996), pp. 337–343
18. Yang, C., Ciesielski, M.: BDS: A BDD-Based Logic Optimization System. IEEE Trans. Comput. Aided Des. Integr. Circ. Syst. **21**(7), 866–876 (2002)

Translation Techniques for Reversible Circuit Synthesis with Positive and Negative Controls

D. Michael Miller and Gerhard W. Dueck

1 Introduction

A reversible Boolean function is a multiple-output function that maps each input assignment to a unique output assignment. Such a function must have the same number of inputs and outputs and the function always has an inverse. The reversible circuit synthesis problem is to realize such a function as a cascade of reversible gates. In this chapter we present function translations that can improve the synthesized circuit making effective use of both positive and negative controls for the reversible gates.

The first function translation considered negates selected function inputs and the corresponding function outputs. If we synthesize a circuit for the translated function, that circuit is easily translated to become a circuit for the original function by changing the polarity of certain controls in the circuit. For an n-input, n-output function, there are 2^n choices for which input-output pairs to negate, hence a broad range of potential circuits for the original function. The case where all input-output pairs are negated translates the original function to its dual.

The second function translation considered permutes input-output pairs. For an n-input, n-output function, there are $n!$ permutations. Note that as the same permutation is applied to the inputs and outputs of the function, if we find a circuit for the translated function, it can be mapped to a circuit for the original function by a simple relabeling of the inputs and corresponding outputs. Swap gates are not

---

D. M. Miller (✉)
University of Victoria, Victoria, BC, Canada
e-mail: mmiller@uvic.ca

G. W. Dueck
University of New Brunswick, Fredericton, NB, Canada
e-mail: gdueck@unb.ca

R. Drechsler, D. Große (eds.), *Recent Findings in Boolean Techniques*,
https://doi.org/10.1007/978-3-030-68071-8_7

needed as is the case in some earlier work where permuting the inputs and outputs are considered as separate operations.

The two function translations can be combined and it is also possible to synthesize a circuit for a function from that function or its inverse. This gives a possibility of $2 \times n! \times 2^n$ translations for a given function. We use transformation-based synthesis techniques to demonstrate the effectiveness of the techniques, but we note that the function translations can be applied with any reversible circuit synthesis method.

Transformation-based synthesis for reversible functions was introduced in 2003 [9]. It is a simple technique that in its most basic form generates a reversible circuit by mapping a given function to the identity by considering the rows of a truth table in order from row 0 to row $2^n - 1$. Variants to that basic approach have been developed and are outlined later in this chapter. A bounded search transformation-based synthesis method is also considered. Results presented show it can be effective but at high computational cost.

A second facet of this chapter is the consideration of ways to simplify reversible circuits with positive and negative controls. To that end, we employ simplification rules presented by Rahman and Rice in [15]. In addition, we consider the use of a generalized form of Peres and inverse Peres gates [14] that allows for both negative and positive controls. Once again it is worth noting that the simplification techniques discussed are applicable to the circuits and are not specifically for transformation-based synthesis. They can be used in conjunction with any other reversible circuit synthesis approach.

Often the goal is to map a reversible circuit to a quantum gate implementation. Here we consider mapping to the NCV gate library [13]. In particular, we present an approach to dealing with negative control CNOT gates which are often not permitted in quantum circuit technologies.

This chapter concludes with an assessment of the positives and limitations of this work with suggestions for ongoing research. The use of negative controls in reversible and quantum circuits, *white dots* as they are often called due to the commonly used graphic, have been considered by a number of researchers [2, 10, 11, 15, 16, 21]. We refer the interested reader to those works and acknowledge them as providing motivation for this work.

2 Background

We here present the necessary background on reversible functions, gates, and circuits as well as necessary detail on NCV quantum circuits. Readers seeking more detailed information should consult [13].

Table 1 A 3 × 3 reversible function

x_2	x_1	x_0	x_2^+	x_1^+	x_0^+
0	0	0	1	1	1
0	0	1	0	0	1
0	1	0	1	0	0
0	1	1	0	1	1
1	0	0	0	0	0
1	0	1	0	1	0
1	1	0	1	1	0
1	1	1	1	0	1

2.1 Reversible Functions, Gates, and Circuits

Definition 1 An n-input, n-output, totally-specified Boolean function $f(X)$, $X = \{x_0, x_1, \ldots, x_{n-1}\}$ is *reversible* if it maps each input assignment to a unique output assignment.

A reversible function can be written as a standard truth table as in Table 1 where + denotes output. The function can also be viewed as a bijective mapping of the set of integers $0, 1, \ldots, 2^n - 1$ onto itself. Hence a reversible function can be defined as an ordered set of integers corresponding to the right side of the table, e.g., $\{7, 1, 4, 3, 0, 2, 6, 5\}$, for the function in Table 1 where the decimal number corresponds to the binary sequence in the obvious way. A reversible function is a permutation and can be expressed as a set of disjoint cycles as done in [19], but we do not follow that approach here.

Definition 2 An m-input, m-output gate is a *reversible gate* if it realizes a reversible function.

In this work, we use the family of mixed-polarity multiple-control Toffoli gates described in Definition 3.

Definition 3 An $m \times m$ *mixed-polarity multiple-control Toffoli (MPMCT) gate* has a single target line and $m - 1$ control lines. Each control is either positive, i.e., activated by a 1, or negative, i.e., activated by a 0. The value on the target line is inverted if all positive controls have value 1 and all negative controls have value 0. The controls are always passed through the gate unaltered.

We write an $m \times m$ MPMCT gate as $T(controls, target)$ where negative controls are indicated by an overline. For example, $T(x_1, \overline{x_2}, x_0)$ denotes an MPMCT gate which inverts the value of x_0 if $x_1 = 1$ and $x_2 = 0$. For drawing gates, $\oplus$ denotes a target, • denotes a positive control, and ○ denotes a negative control.

An MPMCT gate with no controls always inverts the target and is thus the well-known NOT gate. An MPMCT gate with a single control is referred to as a *controlled NOT* (CNOT) and is also known as a Feynman gate [4] if the control is positive. An MPMCT gate with two positive controls is the gate originally proposed by Toffoli [25].

Peres gates and their inverse [14] are often used in reversible circuit synthesis. Conventionally, such a gate has two positive controls. Generalizations of Peres gates have been considered in [11, 24]. In this work we employ mixed-polarity Peres and inverse Peres gates as described in the following definition.

Definition 4 A *mixed-polarity Peres* (MPP) gate is a single gate equivalent to a 2-input mixed-control Toffoli gate followed immediately by a positive-control CNOT gate whose target and control are the controls of the Toffoli gate. A *mixed-polarity inverse Peres* (MPIP) gate is similar except the CNOT immediately precedes the Toffoli. Note these definitions are extensions to the original Peres and inverse Peres gates in that the Toffoli gate can have negative as well as positive controls.

An MPP gate will be denoted $P(c, t_1, t_2)$ where t_1 is the target of the CNOT gate with control c and t_2 is the target of the Toffoli gate with controls c and t_1. t_1 can have an overline to indicate it is a negative when used as a control. An MPIP gate is denoted in the same way with IP instead of P.

Definition 5 An $n \times n$ *reversible circuit* is a cascade of reversible gates with no fanout or feedback.

For example, the circuit in Fig. 1 realizes the function in Table 1. Note the third gate $T(x_2, x_1)$ and the fourth gate $T(x_1, x_2, x_0)$ can be replaced by the inverse Peres gate $IP(x_2, x_1, x_0)$.

2.2 Quantum Gates and Circuits

Reversible circuits can be realized in a variety of technologies. Here we consider the NCV quantum library consisting of four elementary gates: NOT, CNOT, controlled-V, and controlled-$V^\dagger$. We here provide the basic background required to understand the use of this library for the work in this chapter.

Definition 6 The basic information unit in a quantum circuit is the *qubit* whose value is given by $\alpha\,|0\rangle + \beta\,|1\rangle$ where α and β are complex numbers such that $|\alpha|^2 + |\beta|^2 = 1$ and $|0\rangle$ and $|1\rangle$ are basis states normally associated to the Boolean values 0 and 1.

The discussion here is greatly simplified by the fact we are implementing Boolean reversible functions and because the quantum circuits we consider are

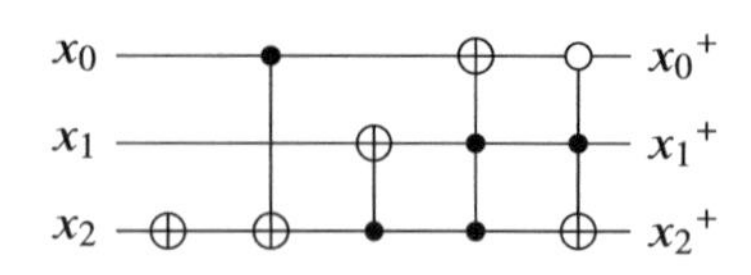

Fig. 1 Reversible circuit for function in Table 1

semi-classical [29] meaning control values are always 0 or 1 thereby avoiding entanglement between qubits.

The operation of a NOT can be expressed as a matrix

$$N = \begin{bmatrix} 0 & 1 \\ 1 & 0 \end{bmatrix}$$

and applying a NOT to a qubit is given by $N \begin{bmatrix} \alpha \\ \beta \end{bmatrix}$. One can see that if the qubit is in a basis state this operation, as expected, flips it to the other basis state.

The matrices defining the V and $V^\dagger$ operations are

$$V = \frac{1}{2}\begin{bmatrix} 1+i & 1-i \\ 1-i & 1+i \end{bmatrix} \quad V^\dagger = \frac{1}{2}\begin{bmatrix} 1-i & 1+i \\ 1+i & 1-i \end{bmatrix}$$

It is readily verified that $N = VV = V^\dagger V^\dagger$. For that reason, V and $V^\dagger$ are called the square roots of NOT. It is also readily verified that $VV^\dagger = V^\dagger V = I$, i.e., V and $V^\dagger$ are the inverses of each other.

V and $V^\dagger$ gates always have a single positive control. When used as a quantum elementary gate, CNOT can also only have a positive control which is different from our use of CNOT in a reversible circuit where we allow a positive or a negative control.

Definition 7 An *ancillary line* is a quantum circuit line used in the realization of an MPMCT gate that is not a control or target for that gate. The value of the ancillary is restored to its value, so operation of the gate effectively has no effect on an ancillary,

Quantum circuit cost is discussed in Sect. 6.

3 Function Translations

A number of reversible circuit synthesis methods have been proposed in [17]. Most employ heuristics and are not guaranteed to find an optimal solution, so it is useful to explore alternative formulations of the synthesis problem and translation of the function to be synthesized in particular.

3.1 Function Inverse

Since a reversible function maps each input assignment to a unique output assignment, such a function has an inverse. The following result is well known [9]:

Theorem 1 *Given a reversible circuit* $g_0, g_1, \ldots, g_{k-1}$ *realizing the reversible function* $f(X)$*, the circuit* $g_{k-1}^{-1}, g_{k-2}^{-1}, \ldots, g_0^{-1}$ *realizes the inverse function* $f^{-1}(X)$.

Proof A reversible gate can be represented by a permutation matrix and a reversible circuit is the product of the matrices for the gates in the circuit. The result follows from the fact that the inverse of a product of matrices is the product of the inverses of the matrices in reverse order. □

This theorem is in fact simpler for circuits composed of MPMCT gates since those gates are all self-inverse, so one need only reverse the order of the gates. Peres gates must be replaced by inverse Peres gates and vice versa.

Given this result, one can synthesize circuits for $f(X)$ and for $f^{-1}(X)$ and use the better of the two as a realization for $f(X)$ where Theorem 1 is applied if the circuit found for $f^{-1}(X)$ is used.

3.2 *Input-Output Negation*

The concept of the dual of a Boolean function is readily extended to reversible functions as per the following definition:

Definition 8 The *dual* of a reversible function $f(X)$ is given by $f^D(X) = \overline{f}(\overline{X})$, where $\overline{f}$ denotes negation of each of the outputs of f and $\overline{X}$ denotes negation of each of the variables in X.

We now show how employing $f^D(X)$ gives a further option for exploring circuits to realize $f(X)$ when both positive and negative gate controls are used.

Theorem 2 *Given an MPMCT circuit G realizing* $f^D(X)$*, a circuit realizing* $f(X)$ *is found by changing all 0 controls to 1 controls and all 1 controls to 0 controls for each gate in G.*

Proof Given an MPMCT circuit for $f^D(X)$, add an inverter to each input and to each output. The result is a circuit realizing $f(X)$ since $f(X) = \overline{f^D}(\overline{X})$. Now move the input inverter from each input across the circuit. As it crosses a control, it inverts that control, and as it crosses a target or passes over a gate not involving the circuit line, it does nothing. When it has passed all gates, it cancels with the corresponding output inverter. So in fact, given an MPMCT circuit G realizing $f^D(X)$, a circuit realizing $f(X)$ is found by simply changing all 0 controls to 1 controls and all 1 controls to 0 controls for each gate in G. □

Definition 8 and Theorem 2 are a special case of a more general translation as follows:

Definition 9 Consider an n-tuple $\alpha = \{\alpha_0, \alpha_1, \ldots, \alpha_{n-1}\}$ where each α_i is either the unary function NOT or the unary identity function. The *α-translation* of a reversible function $f(X)$ is given by $\hat{f}(X) = \tilde{f}(\tilde{X})$ where $\tilde{f}$ denotes application of each α_i to output f_i and $\tilde{X}$ denotes application of each α_i to input x_i.

Theorem 3 *Given an MPMCT circuit G realizing $\hat{f}(X)$, a circuit realizing $f(X)$ is found by changing all 0 controls to 1 controls and all 1 controls to 0 controls for each gate on the lines in G corresponding to those α_i that are NOTs.*

Proof The proof is essentially the proof for Theorem 2 restricted to the lines for which α_i is NOT. □

Given a function f with n variables specified as a truth table, the steps to synthesize a circuit employing Theorem 3 are as follows:

1. Choose an $\alpha = \{\alpha_0, \alpha_1, \ldots, \alpha_{n-1}\}$. The dual is the case where all α_i are NOT.
2. Form a new function $\hat{f}(X)$ as given by Definition 9.
3. Use the chosen synthesis method to find a circuit G for $\hat{f}(X)$ with no MPP or MPIP gate substitutions.
4. Invert all controls for all gates in G on lines for which α_i is NOT.
5. Do any possible MPP and MPIP gate substitutions. The result is a circuit for f.

It is important to note that MPP and MPIP gate substitutions are not performed when finding a circuit for $\hat{f}(X)$ since inverting the controls in that circuit does not properly handle the control polarity associated with the CNOT that was used to form the MPP or MPIP gate.

3.3 *Input-Output Permutation*

Definition 10 An *input-output permutation* is a single permutation σ applied to both the inputs and outputs of a reversible function $f(X)$ yielding a new function $\mathring{f} = \sigma f(\sigma X)$. □

Note that as the same permutation is applied to the inputs and outputs of the function, if we find a circuit for $\mathring{f}(X)$, it can be mapped to a circuit for $f(X)$ by simply reordering the lines in the circuit using σ^{-1}. Swap gates are not needed as is the case in some earlier work where permuting the inputs and outputs are considered as separate operations.

Combining function inverse, input-output negation, and input-output permutation, there are $2 \times 2^n \times n!$ translations of a given reversible function. It is straightforward to translate a circuit for any one of those translations to a circuit for the original function.

4 Transformation-Based Synthesis

As noted earlier, we will use transformation-based synthesis as a means to evaluate the effectiveness of the function translations introduced in the previous section. For ease of description, we present transformation-based synthesis in terms of the truth table representation of a reversible function. Note that transformation-based methods can be implemented using alternate more efficient representations such as decision diagrams [23, 27].

The procedure Map(y, x) described in Algorithm 1 is taken from [22]. It is central to all the transformation-based synthesis algorithms described below. Map identifies a sequence of positive control MCT gates to map the bit pattern y to x where $y > x$. The gates are selected so that they have no effect on any bit pattern $z < x$.

Algorithm 1 begins by setting the control specification c to have as few 1's as possible from y such that $c \geq x$. The latter condition is required to be sure the gates will not affect earlier rows in the truth table. The first **for** loop generates MCT gates with controls c with one gate for each variable outside c that has to be flipped to make y match x. The second **for** loop then uses x as the control and generates one gate for each variable in c that has to be made 0 to match x. In both loops, each gate generated has as its target one of the variables whose value needs to be changed.

Algorithm 1 MCT gate selection to map y to x where $y \geq x$

1: **procedure** MAP(y, x)
2: $glist = empty$
3: **if** $x \equiv y$ **then**
4: return $glist$
5: **end if**
6: $c = y$
7: remove 1 bits from right of c while $c \geq x$
8: $p = (x \oplus y)\&(\sim c)$
9: **for** each bit position $j = 1$ in p **do**
10: g=T(c,j)
11: append g to the end of $glist$
12: **end for**
13: $q = c\&(\sim x)$
14: $c = x$
15: **for** each bit position $j = 1$ in q **do**
16: g=T(c,j)
17: append g to the end of $glist$
18: **end for**
19: return $glist$
20: **end procedure**

Note: T(c,j) denotes a Toffoli gate with positive controls corresponding to 1 bits in c and target j

Basic Algorithm [9] Given a truth table representing a reversible function f, the basic transformation-based synthesis algorithm [9] proceeds through the truth table rows in order $0 \le i < 2^n - 1$. At each row i, if $f(i) \ne i$ MCT gates are selected to map $f(i)$ to i. These gates are chosen such that they do not affect any row j for $j < i$, i.e., those that have already been considered. The gates are added to the circuit being constructed from the output towards the input and the reversible specification is updated by applying the gates to the output side of the specification. When all rows $0 \le i < 2^n - 1$ have been considered, the resulting truth table is the identity function and the gates chosen represent an implementation of the original reversible function. Note that row $2^n - 1$ does not have to be considered as $f(2^n - 1) = 2^n - 1$ when all previous rows match.

Bidirectional Algorithm [9] The bidirectional transformation-based synthesis is a straightforward extension of the basic algorithm. For each row i, the gates G_0 required to transform the output pattern $f(i)$ to i are determined as in the basic algorithm. In addition, there must be a row j later in the table where $f(j) = i$. MCT gates G_1 that transform j to i are determined. The less expensive of G_0 and G_1 is determined and those gates are added to the circuit and used to update f. Note that if G_1 is chosen, the gates apply from the input toward the output of the circuit and are used to update the input side of the specification. The cost of a set of gates can be simply the MCT gate count or can be based on the quantum cost of implementing the MCT gates.

Multi-directional Algorithm [22] In the multi-directional algorithm for each row i every row k, $i \le k \le 2^n - 1$, is considered by mapping both the input and the output patterns to i thereby potentially adding gates to both the input and the output side of the circuit. The algorithm chooses the row k where the mapping has the lowest quantum cost and in the case of a tie the first row k where the mapping results in a function closest to the identity. To see that this algorithm subsumes the previous two note that the basic algorithm is simply the case of only considering row i, while the bidirectional algorithm is the case of only considering two cases: row i and row j where $f(j) = i$.

Search Algorithm It is interesting to consider whether searching can improve upon the above methods. The method given in Algorithm 2 is a simple branch-and-bound search based on the idea behind the multi-directional algorithm.

Search is a recursive procedure with parameters: f the function under consideration, k the row in the truth table of f under consideration, and $glist$ the circuit (list of gates) so far. For the initial call to Search, f should be the function to be realized, $k = 0$, and $glist$ should be empty. The cost of a circuit can be the number of gates or its quantum cost as discussed in Sect. 6. The cost of the best circuit found to date is used to bound the search. $BestCircuit$ and $BestCost$ are globals. Before starting a search, we use the multi-directional algorithm to find the initial $BestCircuit$ and $BestCost$. This is more efficient than setting the initial cost estimate to ∞. Note that the algorithm is presented with some obviously redundant computation for clarity. The actual implementation is more efficient.

Lines 2–4 skip rows in the truth table of f that are already in identity form, i.e., $f_k = k$. Lines 5–8 check if k has reached the end of f which must thus be the

Algorithm 2 Transformation-based search method

```
1: procedure SEARCH(f, k, glist)
2:     while k < N − 1 and f_k = k do
3:         k ← k + 1
4:     end while
5:     if k = 2^n − 1 then
6:         if cost of glist < BestCost then
7:             record glist as BestCircuit and its cost as BestCost
8:         end if
9:     else
10:        for k ≤ j ≤ 2^n − 1 do
11:            G_in ← map(j, k)
12:            G_out ← map(f_j, k)
13:            apply G_in and G_out to map f to g
14:            dist[j] ← Δ(g)
15:        end for
16:        min ← minimum value in dist
17:        if cost(glist) + min * α_0 < BestCost then
18:            for k ≤ j ≤ 2^n − 1 do
19:                if dist[j] ≤ min * α_1 then
20:                    G_in ← map(j, k)
21:                    G_out ← map(f_j, k)
22:                    apply G_in and G_out to map f to g
23:                    Search(g, k + 1, G_in||glist||reverse(G_out))
24:                end if
25:            end for
26:        end if
27:    end if
28: end procedure
```

identity and $glist$ is a completed circuit for the original function. If it is less costly than the best circuit found to date, it is recorded as the best circuit.

Lines 10–16 consider each of the rows j from k through $2^n - 1$ where n is the number of function variables. In each case, the input side j and output side f_j are mapped to k and the gates are applied to map f to a resulting function g. Gates in G_{in} are applied to the input side of f and gates in G_{out} are applied to the output side of f. The idea of trying all $j, k \leq j \leq 2^n - 1$, is carried over from the multi-directional method. For each g, the operator Δ computes the Hamming distance from g to the identity function which is the sum of the Hamming distances between r and g_r for each row of g. The minimum distance is recorded in min.

Line 17 selects whether to continue based on the formula $cost(glist) + min * \alpha_0 < BestCost$ which is estimating the cost of finishing the current circuit based on its cost to date and the minimum Hamming distance found. The factor α_0 is discussed below.

If line 17 determines continuation, lines 18–25 go back through the rows from k to $2^n - 1$. For each, if $dist[j] \leq min * \alpha_1$, g is computed (the factor α_1 is discussed below) and a recursive call is made to Search with parameters $g, k + 1$ (the next row to consider) and the circuit to date which is $glist$ with the gates from G_{in} prepended to the front and the gates from G_{out} reversed and appended to the end. The latter set of gates are reversed because procedure map generates gates from output towards the input when considering mapping an output pattern.

The factor α_0 controls the weight min is giving in estimating the cost of the final circuit. By experiment using NCV quantum cost, we have found 1.33 to be a good value. α_1 in line 19 determines how far $dist[j]$ can be above the minimum for row j to be considered as a basis for further searching. Again by experiment, we have determined that $\alpha_1 = 2$ is effective. More experiments with the search procedure may well lead to a better understanding of the best values for α_0 and α_1 and their interaction.

5 Simplifying a Reversible Circuit

We employ the following simplification rules for MPMCT gates developed by Rahman and Rice [15]. Note that these rules are referred to as templates in [15], but we choose not to call them that to avoid confusion with other formulations of templates in the reversible and quantum circuit literature. Note that rule 1 is the special case of rule 3 with $C = \phi$.

Rahman and Rice Simplification Rules

1. $T(x_c, x_t)T(\overline{x_c}, x_t) = T(x_t) = NOT(x_t)$
2. $T(C, x_t)T(C, x_t) = I$
3. $T(C \cup x_i, x_t)T(C \cup \overline{x_i}, x_t) = T(C, x_t)$
4. (a) $T(C \cup x_i \cup \overline{x_j}, x_t)T(C \cup \overline{x_i} \cup x_j, x_t) = T(x_i, x_j)T(C \cup x_j, x_t)T(x_i, x_j)$
   (b) $T(C \cup \overline{x_i} \cup \overline{x_j}, x_t)T(C \cup x_i \cup x_j, x_t) = T(x_i, x_j)T(C \cup \overline{x_j}, x_t)T(x_i, x_j)$
5. (a) $T(C \cup x_i, x_t)T(C \cup x_j, x_t) = T(x_i, x_j)T(C \cup x_j, x_t)T(x_i, x_j)$
   (b) $T(C \cup \overline{x_i}, x_t)T(C \cup \overline{x_j}, x_t) = T(x_i, x_j)T(C \cup x_j, x_t)T(x_i, x_j)$
   (c) $T(C \cup \overline{x_i}, x_t)T(C \cup x_j, x_t) = T(x_i, x_j)T(C \cup \overline{x_j}, x_t)T(x_i, x_j)$
6. (a) $T(C, x_t)T(C \cup x_i, x_t) = T(C \cup \overline{x_i}, x_t)$
   (b) $T(C, x_t)T(C \cup \overline{x_i}, x_t) = T(C \cup x_i, x_t)$
7. (a) $T(C \cup x_i \cup x_j, x_t)T(C \cup x_k, x_t) = T(x_i \cup x_j, x_k)T(C \cup x_k, x_t)T(x_i \cup x_j, x_k)$
   (b) $T(C \cup x_i \cup x_j, x_t)T(C \cup \overline{x_k}, x_t) = T(x_i \cup x_j, x_k)T(C \cup \overline{x_k}, x_t)T(x_i \cup x_j, x_k)$
   (c) $T(C \cup \overline{x_i} \cup x_j, x_t)T(C \cup x_k, x_t) = T(\overline{x_i} \cup x_j, x_k)T(C \cup x_k, x_t)T(\overline{x_i} \cup x_j, x_t)$
   (d) $T(C \cup \overline{x_i} \cup x_j, x_t)T(C \cup \overline{x_k}, x_t) = T(\overline{x_i} \cup x_j, x_k)T(C \cup \overline{x_k}, x_t)T(\overline{x_i} \cup x_j, x_k)$
   (e) $T(C \cup \overline{x_i} \cup \overline{x_j}, x_t)T(C \cup x_k, x_t) = T(\overline{x_i} \cup \overline{x_j}, x_k)T(C \cup x_k, x_t)T(\overline{x_i} \cup \overline{x_j}, x_k)$
   (f) $T(C \cup \overline{x_i} \cup \overline{x_j}, x_t)T(C \cup \overline{x_k}, x_t) = T(\overline{x_i} \cup \overline{x_j}, x_k)T(C \cup \overline{x_k}, x_t)T(\overline{x_i} \cup \overline{x_j}, x_k)$

For each of the above rules, the substitution holds even if the order of the gates on the left hand side is reversed because they have a common target.

To apply the above simplification rules, we need to be able to determine if two gates can be moved to be adjacent if they are not already. Since a reversible circuit is a cascade of gates, the key operation is to determine if two adjacent gates can be interchanged since moving gates is in fact a sequence of gate interchanges. Since our circuits have both positive and negative controls, checking whether two adjacent gates can be interchanged is more involved then the commonly used so-called moving rule [9].

Moving Rule for MPMCT Gates
Given two adjacent gates the following checks are applied in order:

1. If the two gates have a common control which is positive for one gate and negative for the other, the gates can be interchanged.
2. If the target and controls for one gate all serve as controls for the second gate, in which case the common controls must have equal polarities or (1) would have applied, the gates can be interchanged with the control for the second gate corresponding to the target of the first gate having negated polarity.
3. If the target of one gate is a control for the second, the gates cannot be interchanged, otherwise they can be interchanged.

6 Mapping a Reversible Circuit to a Quantum Circuit

The first step in mapping a reversible circuit to a quantum circuit is to replace each reversible gate by an implementation of that gate comprised of elementary quantum gates, NCV gates in this work. A NOT gate is both a reversible and an elementary quantum gate, so no substitution is required. The same is true for a CNOT with a positive control.

6.1 Negative Control CNOTs

A CNOT with a negative control is not an elementary quantum gate. Two possible substitutions are shown in Fig. 2. Substitution (a) has been used in earlier work. Here we use substitution (b) as only a single NOT needs to be added and it can in fact be placed on either side of the CNOT giving more flexibility for later simplification.

The situation here is complicated by the use of Theorem 3. If a circuit is being synthesized with the intent that the polarity for all gate controls will be flipped, then we want to avoid CNOTs with positive controls. We thus have the notion of a target CNOT control polarity when doing a circuit simplification.

Our procedure for dealing with a CNOT with incorrect control polarity is straightforward. The following steps are applied for each CNOT g_i with incorrect polarity control x_j.

1. We scan from g_i back towards the input to find a $NOT(x_j)$.

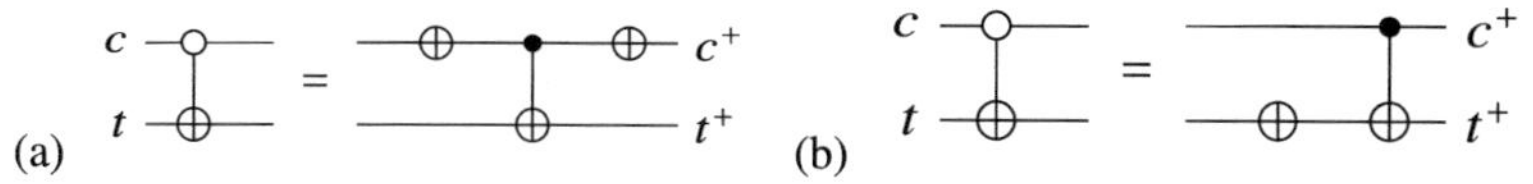

Fig. 2 Mapping a CNOT with a negative control to NCV gates

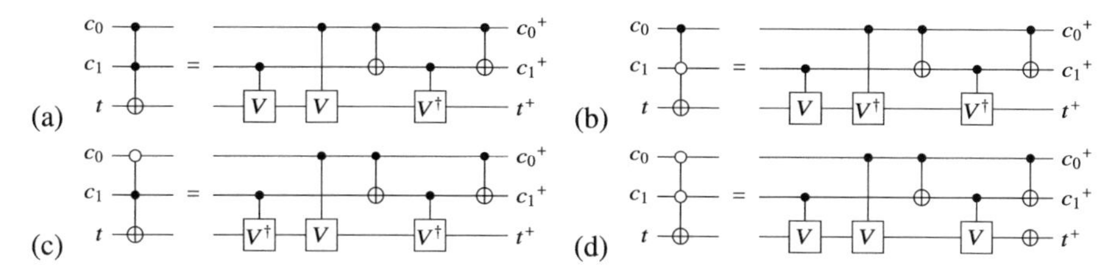

Fig. 3 NCV realization of Toffoli gates dependant on number and placement of negative controls

2. If none is found, we scan from g_i towards the output to find the required $NOT(x_j)$.
3. If a $NOT(x_j)$ is found, in either direction, it is moved across the circuit towards g_i inverting all controls it crosses until it has crossed over g_i.
4. If no $NOT(x_j)$ was found in 1 or 2, a NOT is inserted on the target line of g_i just before g_i and the polarity of the control for g_i is flipped.

6.2 *NCV Realization of MPMCT Gates*

For an MPMCT gate with two controls, the quantum implementation depends on the number and placement of negative controls as shown in Fig. 3. The difference between (b) and (c) is which of the controls is negative. Note that this only affects the assignment of V and $V^\dagger$ to the first two gates. For two negative controls, a sixth gate, a NOT, is required on the target line (t) as shown in (d).

Quantum realizations of MPMCT gates have been extensively studied beginning with the seminal paper by Barenco et al. [1]. To estimate quantum costs during our synthesis procedures, we use results from [18], which are given in Table 2. Each entry in the table is the number of elementary NCV quantum operations required to realize a gate with the associated number of controls.

The first three rows of the table for gates with 0, 1, and 2 controls are costed as described above. We are using substitution (b) for negative control CNOT gates. For three or more controls, a decomposition method is given in [18] which basically expresses a MPMCT gate as a network of gates with fewer controls. It is important to note that for three or more controls, the cost figures given in Table 2 assume one ancillary line is available. Reference [18] includes less expensive realizations if more ancillaries are available.

Table 2 NCV costs of MPMCT gates assuming one ancillary is available if needed

Controls	Negative controls								
	0	1	2	3	4	5	6	7	8
0	1								
1	1	2							
2	5	5	6						
3	14	14	16	18					
4	20	20	20	22	24				
5	32	32	32	34	36	38			
6	44	44	44	44	46	48	50		
7	64	64	64	64	66	68	70	72	
8	76	76	76	76	76	78	80	82	84

6.3 MPP and MPIP Gate Cost and Substitution

Consider Fig. 3 again. The control lines to a two-control MPMCT gate can be interchanged. Furthermore, since an MPMCT gate is self-inverse, the quantum circuit realization can be reversed. Now, since each of the circuits in Fig. 3 has a CNOT between the two MPMCT gate controls, it is clear that that gate will cancel a CNOT between c_1 and c_2 that follows it. The same is also true if the CNOT comes first. This is the basis for substituting the gate pair with an MPP or MPIP gate. One can see that an MPP or MPIP gate formed from an MPMCT with 0 or 1 negative controls thus has a cost of 4. If the MPMCT gate has two negative controls, the cost is 5.

MPP and MPIP gate substitution is straightforward. One need only scan the circuit to find an appropriate two input MPMCT and CNOT gate pair that can be moved together using the procedure described above. The two gates are replaced by a MPP or MPIP gate depending on which side the CNOT lies.

6.4 Overall Simplification and Mapping Strategy

The process applied to simplify a circuit has the following steps:

1. Apply Rahman and Rice MPMCT simplifications.
2. Apply the CNOT correction procedure to deal with any CNOTs that have the wrong control polarity.
3. Apply Rahman and Rice MPMCT simplifications.
4. If any changes to the circuit were made in 1–3, reverse the circuit and apply 1–3 to the result.
5. After iterating 1–4 until there are no changes, if the circuit is in reverse orientation, reverse it.
6. If MPP and MPIP gates are to be used, make all possible MPP and MPIP gate substitutions.

7 Experimental Results

We have implemented the methods described above in C and run our experiments on a x64-based PC with an Intel i5 650 processor and 3GB RAM. Tables 3 and 5 show the results for the $8! = 40,320$ 3-variable reversible functions for nine scenarios. The average quantum cost is shown for each scenario as well as the cpu seconds required.

Two versions of the search method are used. Search A has $\alpha_0 = \alpha_1 = 1$ and Search B has $\alpha_0 = 1.33$ and $\alpha_1 = 2$. Raising the α broadens the scope of the search, i.e., more potential circuits are considered, but that of course incurs increased computational cost.

Scenario (a) in Table 3 is presented to serve as a base line. The methods are applied with no function translations and none of the simplifications discussed in Sects. 5 and 6 including no use of MPP or MPIP gates. Results are then shown for (b) adding function inversion, (c) adding circuit simplification, and (d) adding the use of MPP and MPIP gates.

Table 4 shows the incremental improvements of scenarios (b), (c), and (d) compared to the base case (a). It is interesting to note that adding the use of the function inverse, scenario (b), has marginal effect on the search methods whereas adding the use of MPP and MPIP gates, scenario (d), significantly improves both search methods.

The scenarios in Table 5 all use the function inverse, circuit simplification, and MPP and MPIP gates. Scenarios (e), (f), and (g) show the results for adding use of the dual, input-output negation and input-output permutation separately. Scenarios (h) and (i) show the results for using the dual with permutation and input-output negation with permutation. Scenario (i) gives the best results across Tables 3 and 5.

Table 6 shows the improvements offered by each of scenarios (e) to (i) for each of the synthesis methods. For all methods, scenario (i) using input-output negation and input-output permutation gives the most improvement. This is not surprising as that scenario provides the most function translations to explore for each function. It is interesting that the improvement is not as high for the search methods as for the other three methods. That is because the search methods already explore an extensive solution space.

Table 7 is an analysis of scenario (e) in Table 5 and shows the number of functions for which each method gives the best result among the five synthesis methods. Search B exhibits the best performance but there are exceptions. Separate analysis shows that Search A finds a cheaper result than Search B for 5.1% of the functions. There are even 354 functions for which the Basic method finds the cheapest circuit. These anomalies are due to the heuristic nature of the five methods. Similar results are found for analyses of scenarios (f) to (i).

For the Search B method under scenario (e), 45.96% of the best circuits were found by synthesizing a circuit for $f(X)$, 32.22% by synthesizing a circuit for $f^{-1}(X)$, 11.71% by synthesizing a circuit for the dual of $f(X)$, and 10.11% by synthesizing a circuit for the dual of $f^{-1}(X)$. In total, the 40,320 circuits used

Table 3 Average quantum costs for three variable reversible functions: scenarios (a)–(d)

	(a) No inverse or simplification		(b) Inverse no simplification		(c) Inverse simplification		(d) Inverse simplify MPP MPIP gates	
Synthesis method	cost	cpu(s)	cost	cpu(s)	cost	cpu(s)	cost	cpu(s)
Basic	17.87	0.05	15.99	0.08	14.59	0.33	13.50	0.38
Bidirectional	16.55	0.06	15.55	0.09	14.58	0.29	13.28	0.32
Multi-direct.	16.51	0.30	15.35	0.59	14.41	0.77	13.09	0.82
Search A	14.77	4.92	14.51	9.87	13.60	21.55	11.66	12.66
Search B	14.32	34.55	14.25	68.10	13.40	105.42	11.28	43.00

Table 4 Improvements compared to scenario (a)

	Incremental improvement			
Synthesis method	(b)	(c)	(d)	Total
Basic	10.53%	7.83%	6.10%	24.46%
Bidirectional	6.04%	5.86%	7.85%	19.76%
Multi-direct.	7.03%	5.72%	7.97%	20.71%
Search A	1.76%	6.16%	13.13%	21.06%
Search B	0.49%	5.94%	14.80%	21.23%

Table 5 Average quantum costs for three variable reversible functions: scenarios (e)–(i)

Synthesis	(e) Dual		(f) Input-output negation		(g) Permutation		(h) Dual and permutation		(i) Input-output neg. and permutation	
method[a]	cost	cpu(s)	cost	cpu(s)	cost	cpu(s)	cost	cpu(s)	cost	cpu(s)
Basic	13.01	1.14	12.37	4.31	12.04	2.10	11.88	5.76	11.60	26.12
Bidirectional	12.63	1.00	11.97	3.65	11.84	1.76	11.65	4.82	11.31	21.68
Multi-direct.	12.48	1.86	11.85	7.75	11.72	4.78	11.54	10.97	11.21	46.15
Search A	11.49	37.78	11.33	135.16	10.94	74.40	10.90	181.84	10.84	819.35
Search B	11.16	160.04	11.07	648.62	10.76	261.93	10.73	762.61	10.69	3802.06

[a]Function inverse, circuit simplification and MPP and MPIP gates used in all scenarios.

Table 6 Improvements compared to scenario (d)

	Improvements				
Synthesis method	(e)	(f)	(g)	(h)	(i)
Basic	3.63%	8.39%	10.81%	12.00%	14.07%
Bidirectional	4.89%	9.89%	10.84%	12.27%	14.85%
Multi-direct.	4.66%	9.50%	10.47%	11.84%	14.39%
Search A	1.46%	2.84%	6.17%	6.52%	7.08%
Search B	1.06%	1.86%	4.61%	4.88%	5.23%

14,396 MPP and 13,275 MPIP gates. It is interesting to note that of those 27,671 gates, 33.15% had two positive controls, 10.7% had two negative controls, and 56.16% had one positive and one negative control. This demonstrates the usefulness of allowing MPP and MPIP gates rather than just Peres and inverse Peres gates.

Table 7 Best results for scenario (e) in Table 5

Synthesis method	Best result No. of functions	% of total	Unique best result No. of functions	% of total
Basic	13,968	34.64%	354	0.88%
Bidirectional	17,135	42.50%	122	0.30%
Multi-directional	18,349	45.51%	34	0.08%
Search A	30,082	74.61%	1744	4.33%
Search B	37,678	93.45%	8975	22.26%

Table 8 Search B for scenario (i)—distribution by function translation

	Permutation index							
Negation	0	1	2	3	4	No perm.	Subtotal	% of total
No function inversion								
No neg.	332	376	726	1913	2870	19,514	25,731	63.82%
1	17	18	20	56	65	137	313	0.78%
2	16	21	21	63	67	153	341	0.85%
3	14	15	17	65	67	223	401	0.99%
4	11	14	24	72	67	224	412	1.02%
5	15	21	26	86	88	355	591	1.47%
6	17	20	31	94	96	441	699	1.73%
Dual	19	22	33	131	135	1083	1423	3.53%
Subtotal	441	507	898	2480	3455	22,130	29,911	74.18%
% of total	1.09%	1.26%	2.23%	6.15%	8.57%	54.89%		
With function inversion								
No neg.	234	260	368	921	1248	4416	7447	18.47%
1	16	17	19	50	55	112	269	0.67%
2	14	15	19	43	49	115	255	0.63%
3	14	15	17	50	53	184	333	0.83%
4	9	12	19	63	58	174	335	0.83%
5	11	17	17	64	63	245	417	1.03%
6	15	18	25	79	73	284	494	1.23%
Dual	16	19	28	103	93	600	859	2.13%
Subtotal	329	373	512	1373	1692	6130	10,409	25.82%
% of total	0.82%	0.93%	1.27%	3.41%	4.20%	15.20%		

Table 8 shows the distribution of the best circuits found for the 40,320 three variable functions using method Search B for scenario (i) in Table 5.

To place the above results in some context, the authors of [2] considered the realization of three variable reversible function using mixed-polarity Toffoli gates and positive polarity Reed-Muller techniques. The best results they report have an average quantum cost of 13.36 which is better than our base line but higher than our best results. This illustrates further the advantage of using MPP and MPIP gates and the circuit simplification results discussed in this chapter.

Table 9 Selected *worst case* functions

Synthesis method	cost^a	cost^b	cost^c	code	cpu(s)	cost^a	cost^b	cost^c	code	cpu(s)
	3_17					*4_49*				
Basic	15	15	11	P3-I	0.00	80	68	43	P6-N5	0.06
Bidirectional	15	12	11	P3	0.00	101	87	31	P6-N11-I	0.04
Multi-direct.	15	12	11	P3	0.00	69	70	32	P6-N11	0.08
Search A	15	12	11	N3	0.01	55	52	32	P6-N11	3.55
Search B	15	12	11	P4-I	0.06	36	32	28	P17-N11	476.90
Best known cost	10 [28]					32 [28]				
	hwb4					*hwb5*				
Basic	71	53	32	P22-D	0.05	352	294	199	P119-N3-I	4.24
Bidirectional	64	52	21	P22-I	0.04	323	301	171	P116-N20-I	2.84
Multi-direct.	58	55	21	P22-I	0.07	313	282	172	P115-N3-I	5.35
Search A	49	41	21	P22-I	3.39	280	230	101	P110-N16-I	6909.33
Search B	27	21	20	P20-I	12.97	see note *d*				
Best known cost	19 [8]					71 [8]				

[a]Using inverse translation, no circuit simplification or MPP/MPIP gates
[b]Using inverse translation, circuit simplification and MPP/MPIP gates
[c]Using inverse, input-output negation, input-output permutation, circuit simplification and MPP/MPIP gates
[d]Search B for hwb5 is computationally prohibitive

De Vos and Van Rentergem [3] have presented a reversible circuit synthesis approach using Young-based subgroups. They consider circuits with positive and negative controls. A difference from the work here is that the control function for a gate can be any Boolean function not just the conjunction of controls. They allow CNOT gates with a negative control. They did not use Peres type gates. For 3-variable reversible functions, they report average gate counts of 5.88 for their Algorithm A, 4.21 for their Algorithm B and 3.73 as the optimal average. When we apply our methods allowing negative control CNOT gates, using MPP or MPIP gates and using the dual and choice of a circuit for f or f^{-1}, we find gate averages of 4.73 (Basic), 4.57 (Bidirectional), 4.50 (Multi-directional), 4.49 (Search A), and 4.50 (Search B). Note that this is a very rough comparison as we are using Peres type gates and all our gates use a conjunctive control function.

Table 9 shows the results for four functions which have been described as *worst cases* for several synthesis methods [28]. They are certainly known to be difficult for transformation-based synthesis. Note that the results show the cost of the best circuit when considering the synthesis of f and f^{-1} and using input-output negation and input-output permutation. All simplification techniques discussed in Sects. 5 and 6 are applied including the use of MPP and MPIP gates.

In Table 9, the function translation resulting in the best circuit found for each case is shown in column trans. P-x indicates input-output permutation has been applied where x denotes the permutation index as defined in [12]. N-x indicates

Table 10 Search B applied to hwb5 for four function translations

Translation	cost	Solution cpu(s)	Total cpu(s)
Function	120	2538.43	3600.00
Inverse	85	2444.29	2565.57
Dual	144	1503.20	3600.00
Inverse dual	152	614.17	3600.00

$$T(d, e)T(e, d)T(b, e)T(a, b)T(d, b)T(a, d)T(a, e)T(d, e, a)T(a, b)T(c, e, b)P(c, \overline{d}, a)$$
$$T(a, c)T(a, e)T(b, a)T(a, d)T(b, e, a)T(a, b)T(a, d)T(b, d, a)T(a, d)T(b, c, a)T(a, e, d)$$
$$T(a, b)T(\overline{b}, e, c)T(a, \overline{b}, d, e)T(a, c, d)IP(\overline{b}, e, c)T(d, b)T(c, \overline{e}, a)T(a, e)$$

Fig. 4 hwb5 circuit found using Search B: 30 gates, quantum cost 85

input-output negation has been applied, after possible permutation, with the 1's in the binary expansion of x indicating which input-output positions are negated. D is used to indicate all input-output pairs are negated, i.e., the dual is used. Lastly, I indicates the function is inverted following any permutation and negation.

Method Search B is computationally much more expensive than the other methods. For that reason, applying Search B to hwb5 is omitted in Table 9. Instead, Table 10 gives the results for applying Search B to hwb5 for four scenarios: the function, the inverse of the function, the dual of the function, and the dual of the inverse of the function. For each scenario, a limit of 1 h CPU time was imposed.

The best circuit was found using the inverse function. Interestingly that scenario did not hit the 1 h time limit which means the full search was completed. The circuit found using the inverse of hwb_5 is shown in Fig. 4. Note that it uses MPMCT gates with negative controls as well as an MPP and an MPIP gate. This circuit is far better than the results reported for hwb_5 in Table 9 but finding it required very lengthy computation time.

This result is reasonably close to the best circuit found to date, cost 71, which is listed on Maslov's benchmark web site [8]. That circuit was posted by the authors of [26] which presented a variable-length chromosome evolutionary algorithm for reversible circuit synthesis. The cpu usage required to find the circuit is not reported. It is interesting that a relatively simplistic transformation-based synthesis approach can produce so good a circuit especially compared to other techniques (see [8, 28] for examples of circuits for hwb5).

8 Heuristic Selection of Function Translations

The results presented in the previous section show that the use of input-output negation and input-output permutation can significantly reduce circuit cost. However it is clear that searching through all possible translations quickly becomes impractical as n, the number of variables, increases since there are 2^n input-output negations and $n!$ input-output permutations. The ideal would be able to pick a small number

Table 11 Selected translation scenarios for 3-variable functions using the basic method

	Translations	Cost	% Impr. over (a)	cpu(s)
1	Inverse (a)	13.50		0.37
2	Inverse, dual	13.04	3.47%	0.98
3	Inverse, HION	12.80	5.23%	1.42
4	Inverse, full I-O negation	12.37	8.41%	4.24
5	Inverse, HIOP	12.52	7.29%	0.73
6	Inverse, all I-O perm.	12.04	10.86%	2.05
7	Inverse, HION, HIOP	12.11	10.32%	2.91
8	Inverse, all I-O neg., all I-O perm.	11.60	14.09%	25.17

of translations to consider based on properties of the function to be synthesized. As a start toward that goal, we here present two heuristic methods for choosing which translations to consider.

Heuristic Input-Output Negation (HION) We use the following:

1. No input-output negations applied.
2. All input-output pairs negated, i.e., the dual.
3. The α-translation, see Definition 9, where α_i is the identity if $x_i = 0$ and α_i is NOT if $x_i = 1$ in the earliest assignment to $(x_0, x_1, \ldots, x_{n-1})$ where the Hamming distance between that assignment and the corresponding output assignment $(x_0^+, x_1^+, \ldots, x_{n-1}^+)$ is minimal where *earliest* refers to considering input assignments starting from $(0, 0, \ldots, 0)$ in truth table order.

Heuristic Input-Output Permutation (HIOP) We employ two permutations:

1. The inputs and outputs in the order given, i.e., the null permutation.
2. The reverse permutation where the inputs and outputs are in the reverse of the order given, e.g., x_0, x_1, x_2 is permuted to x_2, x_1, x_0.

Table 11 presents results of applying the basic transformation-based synthesis method to the 40,320 3-variable reversible functions for a variety of function translation scenarios. Circuit simplification and MPP and MPIP gates are used in all cases.

The top row of Table 11 uses only the function inverse translation and is intended as a baseline for measuring the improvement offered by the other scenarios. The function inverse translation is employed in all the other scenarios.

Lines 2–4 show the results for using the dual, heuristic input-output negation, and for comparison using all eight possible input-output negations. Lines 5 and 6 compare using heuristic input-output permutation and all six possible input-output permutations. Lastly, rows 7 and 8 compare using both heuristic translation approaches with using all function translations. In all cases, improvement is measured relative to just using the function inversion translation and shows the percentage reduction in the circuit quantum cost.

The results indicate that while our heuristic methods show quite good improvement, they, as one would expect for such simple approaches, fall well short of employing all function translations. Applying the basic synthesis method and heuristic translation approaches to the worst case function 3_17 reduces the quantum cost from 15 down to 11 for the function inverse, reverse permutation, and no input-output negation. However, applying the techniques to the other three worst case functions offers no improvement over just using only the function inverse translation. That is simply the nature of those functions but does suggest that the heuristic techniques described in this section can likely be improved upon.

9 Discussion and Future Work

This chapter has considered a number of aspects of the synthesis and simplification of reversible circuits with positive and negative controls. The main contributions are the function translation techniques of input-output negation and input-output permutation introduced in Sect. 4, the employment of mixed control Peres gates, and various circuit simplifications discussed in Sects. 5 and 6 which combine well with Rahnan and Rice's MPMCT techniques. We plan to investigate whether the circuit rewriting rules proposed in [21] can improve our circuit simplification procedures.

The experimental results presented in this chapter demonstrate that the input-output negation and input-output permutation function translations introduced here have significant potential for improving the synthesis of reversible circuits. However, the approach of searching all possible translations is clearly limited. We have suggested some initial heuristic techniques to choose translations to consider by examining the function to be synthesized. The results show some promise, but it is very likely that incorporating the choice of function translations into the synthesis process will be more effective than making a pre-synthesis choice. This is an area requiring further research.

The transformation-based search method is of potential interest but is truly a work in progress. More work is needed to determine how best to bound the search in order to make it computationally feasible. In addition, the authors of [6, 20] have incorporated Fredkin gates [5], which are controlled swap gates, into transformation-based synthesis. It would be interesting to see how their approaches might be incorporated into our work and how Fredkin gates might be extended to be mixed-polarity multiple control gates.

The work thus far has been limited to considering transformation-based synthesis methods. It would be very interesting to see how effective the function translations input-output negation and input-output permutation are when combined or integrated into other reversible circuit synthesis methods.

This work has shown that mixed polarity Peres and inverse Peres gates can be quite effective in reducing quantum circuit cost. The approach here has been to consider the introduction of such gates into a circuit as part of the circuit simplification process. It would be interesting to consider integration of these gate types into the synthesis process itself both for transformation-based and other synthesis methods.

The comparison of the methods presented in this chapter to the work of De Vos and Van Rentergem [3], despite the differences in gate types permitted, suggests there is room for further improvement of our methods.

The quantum cost used in this chapter is based on the NCV gate library. Future work should consider cost metrics for fault-tolerant quantum circuits using the Clifford+T quantum gate library [13]. Another area to consider would use Qiskit or ProjectQ [7] to map an MCMPT circuit into an NISQ compatible circuit and then count gates after performing optimizations at the quantum gate level.

References

1. Barenco, A., Bennett, C.H., Cleve, R., DiVincenzo, D.P., Margolus, M., Shor, P., Sleator, T., Smolin, J.A., Weinfurter, H.: Elementary gates for quantum computation. Phys. Rev. A **52**(5), 3457–3467 (1995)
2. Cheng, C.S., Singh, A.K., Gopal, L.: Efficient three variables reversible logic synthesis using mixed polarity Toffoli gate. Procedia Comput. Sci. **70**, 362–368 (2015)
3. De Vos, A., Van Rentergem, Y.: Young subgroups for reversible computers. Adv. Math. Commun. **2**, 183–200 (2008)
4. Feynman, R.: Quantum mechanical computers. Optic News (1985), pp. 11–20
5. Fredkin, E., Toffoli, T.: Conservative logic. Int. J. Theor. Phys. **21**, 219–253 (1982)
6. Handique, M., Singha, R.: A modified transformation-template based synthesis using Fredkin/swap gates in reversible circuits. Procedia Comput. Sci. **125**, 801–809 (2018)
7. LaRose, R.: Overview and comparison of gate level quantum software platforms. Quantum **3**, 130 (2019)
8. Maslov, D.: Reversible logic synthesis benchmarks page. http://www.cs.uvic.ca/~dmaslov/
9. Miller, D.M., Maslov, D., Dueck, G.W.: A transformation-based algorithm for reversible logic synthesis. In: Proceedings of IEEE/ACM Design Automation Conference (DAC), pp. 318–323 (2003)
10. Moraga, C.: Using negated control signals in quantum computing circuits. Facta Universitatis(Nis) Ser. Electr. Energy **24**(3), 423–435 (2011)
11. Moraga, C.: Mixed polarity reversible Peres gates. Electron. Lett. **50**(14), 987–989 (2015)
12. Myrvold, W., Ruskey, F.: Ranking and unranking permutations in linear time. Inf. Process. Lett. **79**, 281–284 (2001)
13. Nielsen, M., Chuang, I.: Quantum Computation and Quantum Information. Cambridge University Press, Cambridge (2000)
14. Peres, A.: Reversible logic and quantum computers. Phys. Rev. A **32**(6), 3266–3276 (1985)
15. Rahman, M.Z., Rice, J.E.: Templates for positive and negative control Toffoli networks. In: Lecture Notes in Computer Science, vol. 8507 (2014)
16. Ribeiro, A., Kowada, L., Marquezino, F., Figueiredo, C.: A New reversible circuit synthesis algorithm based on cycle representations of permutations. Electron. Notes Discrete Math. **50**, 187–192 (2015)
17. Saeedi, M., Markov, I.L.: Synthesis and optimization of reversible circuits - A survey. CoRR abs/1110.2574 (2011). http://arxiv.org/abs/1110.2574
18. Sasanian, Z., Miller, D.M.: NCV realization of MCT gates with mixed controls. In: Proceedings of 2011 IEEE Pacific Rim Conference on Communications, Computers and Signal Processing, pp. 567–571 (2011)

19. Shende, V.V., Prasad, A.K., Markov, I.L., Hayes, J.P.: Reversible logic circuit synthesis. In: ICCAD. San Jose, California, USA, pp. 125–132 (2002)
20. Soeken, M., Chattopadhyay, A.: Fredkin-enabled transformation-based reversible logic synthesis. In: Proceedings of the International Symposium on Multiple-Valued Logic, pp. 60–65 (2015)
21. Soeken, M., Thomsen, M.K.: White dots do matter: rewriting reversible logic circuits. In: Lecture Notes in Computer Science, vol. 7948 (2013)
22. Soeken, M., Dueck, G.W., Rahman, M.M., Miller, D.M.: An extension of transformation-based reversible and quantum circuit synthesis. In: Proceedings of the International Symposium on Circuits and Systems, pp. 2290–2293 (2016)
23. Soeken, M., Tague, L., Dueck, G.W., Drechsler, R.: Ancilla-free synthesis of large reversible functions using binary decision diagrams. J. Symb. Comput. **73**, 1–26 (2016)
24. Szyprowski, M., Kerntopf, P.: Low quantum cost realization of generalized Peres and Toffoli gates with multiple-control signals. In: 2013 13th IEEE International Conference on Nanotechnology (IEEE-NANO 2013), pp. 802–807 (2013)
25. Toffoli, T.: Reversible computing. Tech memo MIT/LCS/TM-151, MIT Lab for Comp. Sci (1980)
26. Wang, X., Jiao, L., Li, Y., Qi, Y., Wu, J.: A variable-length chromosome evolutionary algorithm for reversible circuit synthesis. Multiple-valued Logic Soft Comput. **25**, 643–671 (2015)
27. Wille, R., Drechsler, R.: BDD-based synthesis of reversible logic for large functions. In: Proceedings of the Design Automation Conference, pp. 270–275 (2009)
28. Wille, R., Große, D., Teuber, L., Dueck, G.W., Drechsler, R.: RevLib: An online resource for reversible functions and reversible circuits. In: Int'l Symposium on Multi-Valued Logic, pp. 220–225 (2008). RevLib is available at www.revlib.org
29. Yamashita, S., Minato, S., Miller, D.M.: Synthesis of semi-classical quantum circuits. Multiple-Valued Logic Soft Comput. **18**(1), 99–114 (2012)

Hybrid Control of Toffoli and Peres Gates

Claudio Moraga

1 Introduction

Toffoli [9] and Peres [7] gates are basic components of reversible circuits. Figure 1 shows their symbolic representation and functionality following the definitions given by their authors. Characteristic of these representations is the use of dots to signalize the connection between the control signals and the gate to be controlled. Originally the dots were black and defined to represent an effective control signal with value 1. Much later white dots were introduced (see [4, 8]) to represent effective control signals with value 0.

In [5] Toffoli gates with disjunct control were introduced, adding some flexibility to the synthesis of reversible circuits. For this kind of gate, instead of a dot as connecting symbol between a control bit and a target gate, an upside down (black) triangle ▼ was chosen for its similarity with the disjunction symbol $\vee$ used in mathematical logic. The symbol, functionality, its Barenco et al. type of quantum model [2], and its specification matrix is shown in Fig. 2.

It is simple to see in the Barenco et al. type model that if c_1 has the value 1 and c_2 has the value 0, the first V-gate will become active, and the second one will be inhibited and behaving as the identity. (For short, in what follows it will simply be said "if c_1 is 1 and c_2 is 0".) Furthermore, c_1 will activate the $CNOT$ gates, producing a "local 1" that will activate the third V-gate. (The last $CNOT$ gate only recovers the original value of c_2). Finally, the cascade of the two active V-gates produce the expected NOT behavior.

C. Moraga (✉)
Faculty of Computer Science, Technical University of Dortmund, Dortmund, Germany

Department of Informatics, Technical University "Federico Santa María", Valparaíso, Chile
e-mail: claudio.moraga@tu-dortmund.de

R. Drechsler, D. Große (eds.), *Recent Findings in Boolean Techniques*,
https://doi.org/10.1007/978-3-030-68071-8_8

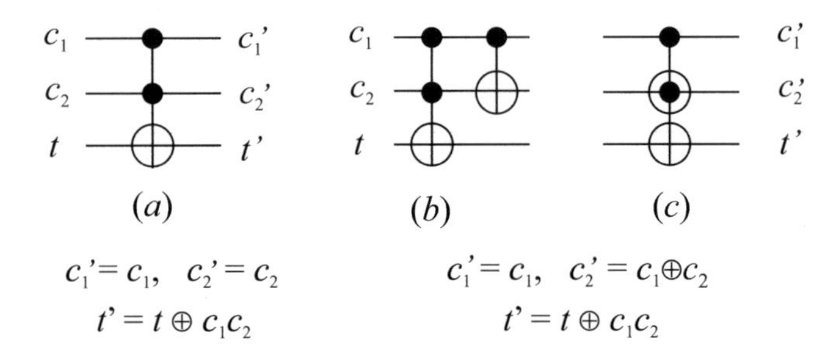

Fig. 1 (**a**) Symbol and functionality of a Toffoli gate. (**b**, **c**) Symbols and functionality of a Peres gate

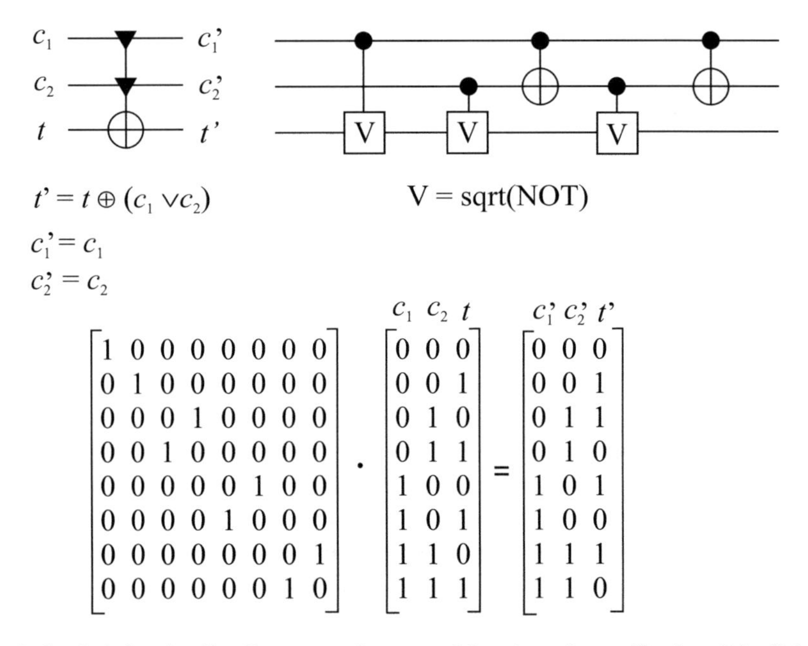

Fig. 2 Symbol, functionality, Barenco et al. type model, and matrix specification of the disjunctively controlled Toffoli gate

In the case that c_1 is 0 and c_2 is 1, the first V-gate and the $CNOT$ gates will be inhibited, whereas the second and the third V-gates will be activated and will produce the expected NOT behavior. Finally, if both c_1 and c_2 are 1, the two first V-gates become activated and produce the expected NOT behavior. Since the $CNOT$ gates will also become activated by c_1, the first one will produce a local 0 by negating c_2, inhibiting the third V-gate.

It is fairly obvious that if both c_1 and c_2 are 0, then all gates will be inhibited.

From the above analysis of cases clearly follows that all three elementary gates of the Barenco et al. type model are equal: either V or $V^\dagger$. Recall that $V^2 = (V^\dagger)^2 =$

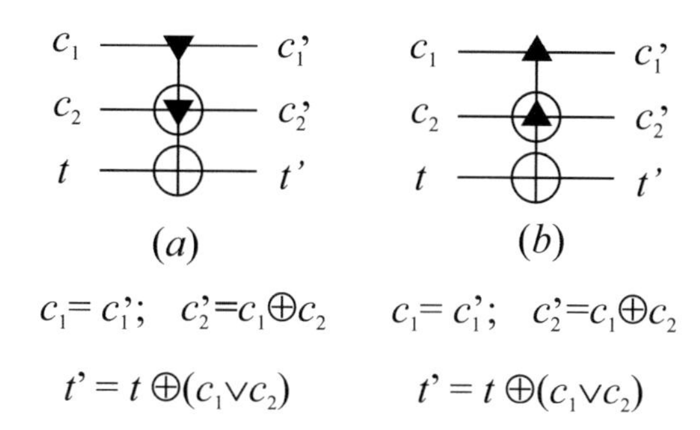

Fig. 3 (**a**) Symbol and functionality of a disjunctively controlled Peres gate. (**b**) Symbol and functionality of a disjunctively controlled inverse Peres gate

NOT. It is interesting to mention that except for [5], no other circuits using disjunct control of reversible gates seem to be available in the literature. However, a Clifford $+T$ circuit [1, 3] with the functionality of a disjunctively controlled Toffoli gate based on $\overline{\bar{c}_1\bar{c}_2}$ appears in [1].

In the conjunctive control case, Peres gates [7] are closely related to the Toffoli gates. At the target level, they have the same functionality but $c_2' = c_1 \oplus c_2$. Moreover, Peres gates are not self-inverses. An inverse Peres gate has been designed such that its Barenco et al. quantum model is the mirror of that of the original Peres gate. These features are preserved for the Peres gate with disjunctive control. This is shown in Fig. 3. The symbol for the "OR-Peres" gate has been made in analogy to that of Fig. 1c.

2 Hybrid-Controlled Gates

In this section, the following question will be considered: Is it possible to have gates with k controls, $k > 2$, (or their efficient building blocks realization based on the Barenco et al. and [5, 6] quantum models), realizing the specifications of Eq. (1)?

$$t' = t \oplus (c_j \vee \bigwedge_{\substack{i=1 \\ i \neq j}}^{k} c_i) \quad \text{or} \quad t' = t \oplus (c_j \wedge \bigvee_{\substack{i=1 \\ i \neq j}}^{k} c_i). \tag{1}$$

A control strategy leading to the above results will be called *hybrid*.

Without loss of generality, the case $k = 3$ and $j = 1$ will be first analyzed. The symbolic representation of these cases is shown in Fig. 4.

A straightforward realization of the gate of Fig. 4a is shown in Fig. 5. The circuit of Fig. 5 has a quantum cost of 13 [2, 5], but it requires an ancillary line driven by 0. Notice that the dotted $CNOT$ gates cancel each other changing the Toffoli gates into a Peres gate and an inverse Peres gate, respectively, thus reducing by 2 the quantum cost.

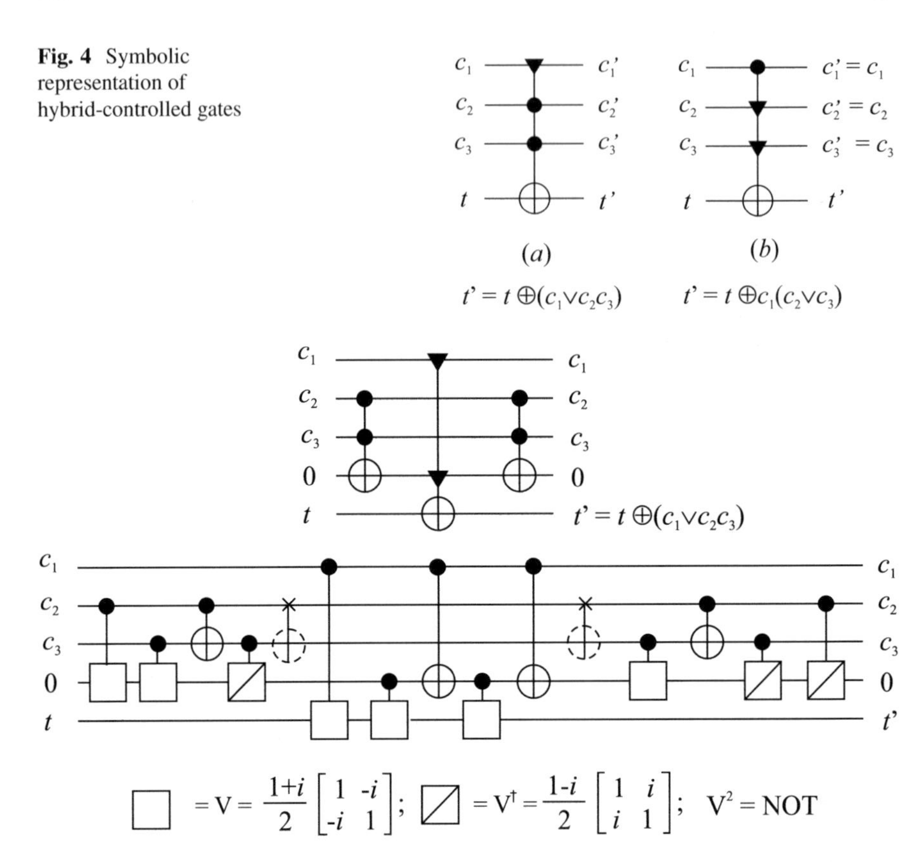

Fig. 4 Symbolic representation of hybrid-controlled gates

Fig. 5 Realization of the gate of Fig. 4a. Symbolic diagram and quantum circuit based on [2] and [5]

It should be recalled that in the context of a quantum realization, an ancillary line is an additional qubit, and this may increase the decoherence risk of the circuit.

A better realization may be obtained based on the equivalence:

$$c_1 \vee c_2c_3 = c_1 \oplus c_2c_3 \oplus c_1c_2c_3. \tag{2}$$

The circuit shown in Fig. 6 is ancilla-free. It has a quantum cost of 19, and it makes use of the quantum gates $CNOT$, $V = sqrt(NOT)$ and $W = sqrt(V)$.

An even better realization may be obtained considering that the following holds:

$$c_1 \vee c_2c_3 = \overline{\overline{c}_1\overline{c_2c_3}} = 1 \oplus \overline{c}_1\overline{c_2c_3} = 1 \oplus \overline{c}_1 \oplus \overline{c}_1c_2c_3 = c_1 \oplus \overline{c}_1c_2c_3. \tag{3}$$

Equation (3) leads to the circuit shown in Fig. 7.

The realization shown in Fig. 7 is ancilla-free and has a quantum cost of 14. It makes use of $CNOT$, W, and $W^\dagger$ gates. It may be seen that the first $CNOT$ gate

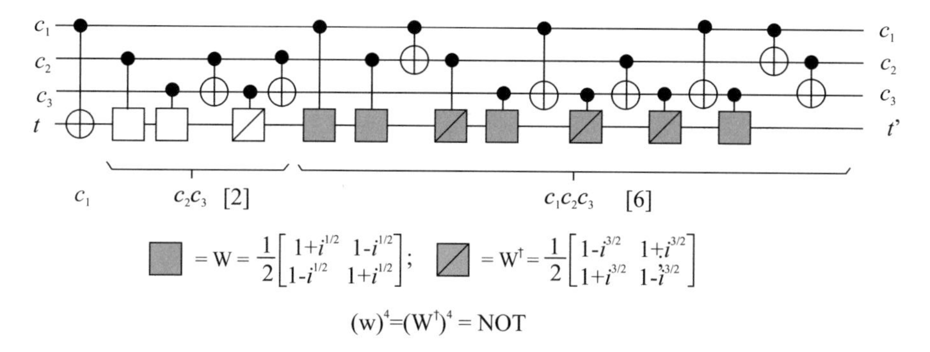

Fig. 6 Ancilla-free quantum circuit for $t \oplus (c_1 \vee c_2c_3)$

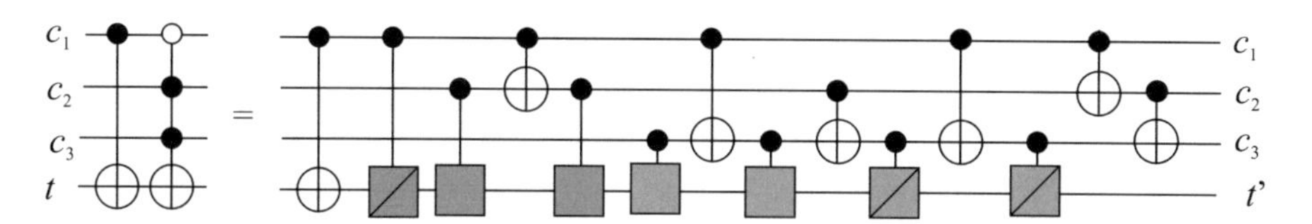

Fig. 7 Quantum realization $t' = t \oplus (c_1 \vee c_2c_3) = t \oplus c_1 \oplus \overline{c}_1c_2c_3$

may be moved to the end of the circuit and could be activated together with the $CNOT$ gate acting on c_3 controlled by c_2. Then the circuit would have a depth of 12. Of all shown circuits, this is the realization with lowest quantum cost and shortest depth.

With respect to the gate of Fig. 4b, a straightforward realization taking advantage of Peres gates, which at the target level have the same functionality as the Toffoli gate, but with a lower quantum cost, is shown in Fig. 8.

The circuits of Fig. 8 have the same structure, quantum cost, and drawback of the circuit shown in Fig. 5, mainly, the use of an ancilla qubit.

For a good realization of a circuit for the gate of Fig. 4b, it is quite reasonable to analyze the dual of Eq. (3):

$$c_1(c_2 \vee c_3) = c_1(\overline{\overline{c}_2\overline{c}_3}) = c_1 \oplus c_1\overline{c}_2\overline{c}_3. \tag{4}$$

Equation (4) leads to the following "dual" circuit:

The circuit of Fig. 9 inherits all properties of the circuit of Fig. 7 with a quantum cost of 14 and a possible depth of 12.

3 Scalability

In this section, equation (1) will be used again, but with $j \neq 1$ and $k > 3$.

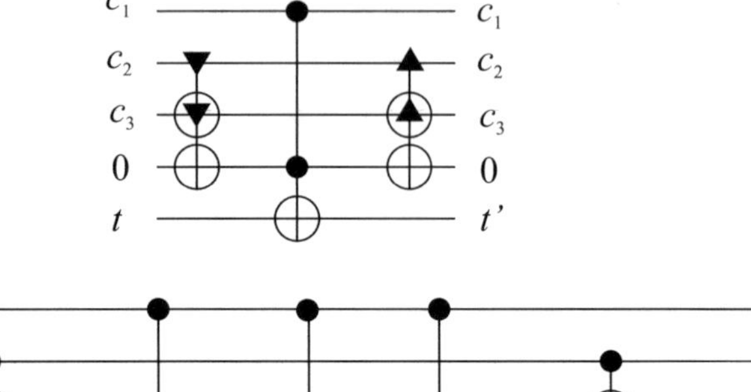

Fig. 8 Realization of the gate of Fig. 4b. Symbolic diagram and quantum circuit based on [5] and [2]

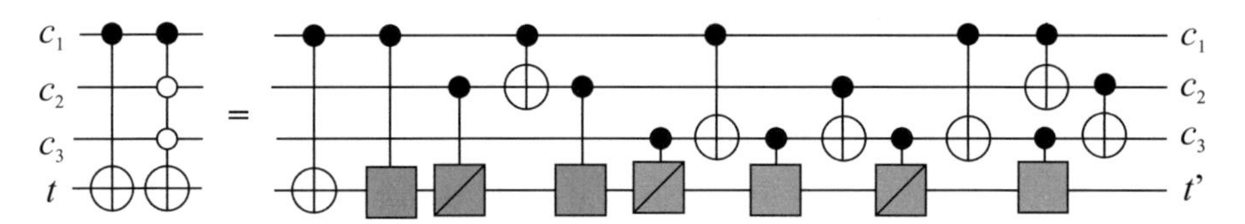

Fig. 9 Quantum realization of the hybrid-controlled gate of Fig. 4b

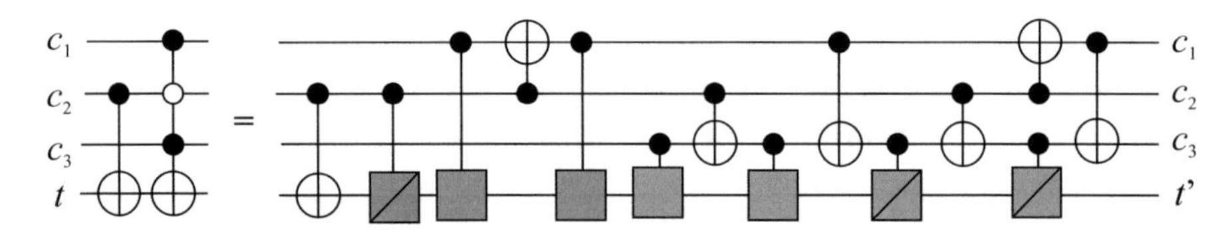

Fig. 10 Quantum realization of $t' = t \oplus (c_2 \vee c_1c_3) = t \oplus c_2 \oplus c_1\overline{c}_2c_3$

If $j = 2$ or 3, and $k = 3$, no swap gates are necessary. Only the functionality of the lines will be accordingly modified. This is shown in Fig. 10 with respect of the circuit of Fig. 7. A similar change would take place in the circuit of Fig. 9.

If $k > 3$ and $j \neq 1$, Eqs. (3) and (4) may be straightforward extended to satisfy the new requirements.

$$
\begin{aligned}
c_j \vee c_1c_2\cdots c_{j-1}c_{j+1}\cdots c_k &= \overline{\overline{c}_j\overline{c_1c_2\cdots c_{j-1}c_{j+1}\cdots c_k}} \\
&= 1 \oplus \overline{c}_j\overline{c_1c_2\cdots c_{j-1}c_{j+1}\cdots c_k} \\
&= 1 \oplus \overline{c}_j \oplus \overline{c}_jc_1c_2\cdots c_{j-1}c_{j+1}\cdots c_k \qquad (5) \\
&= c_j \oplus \overline{c}_jc_1c_2\cdots c_{j-1}c_{j+1}\cdots c_k \\
&= c_j \oplus c_1c_2\cdots c_{j-1}\overline{c}_jc_{j+1}\cdots c_k
\end{aligned}
$$

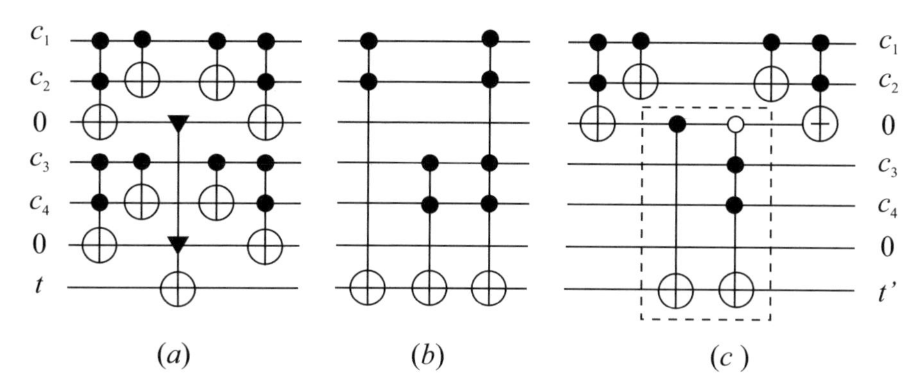

Fig. 11 Circuits for Eqs. (7) and (8). (**a**) Decomposition using Peres and inverse Peres gates. (**b**) $GF(2)$ decomposition. (**c**) Decomposition using Peres gates and the circuit of Fig. 7 (shown in a dotted box)

A similar result is obtained if Eq. (4) is extended. In both cases, the Toffoli gate will increase the number of control signals. The method presented in [6] shows how to obtain the quantum model of Toffoli gates with an increasing number of control lines. The functionality of the lines will possibly have to be reordered, depending on the value of j, as illustrated in Fig. 10.

A next step of scalability considers the sets $N_n = \{1, 2, \ldots, n\}$, $Q \subset N_n$, $R \subset N_n$, $|Q| = q$, $|R| = r$, where $Q \cap R = \Phi$. In analogy to Eq. (1), let

$$t' = t \oplus (\bigwedge_{i \in Q} c_i \vee \bigwedge_{j \in R} c_j). \tag{6}$$

Without loss of generality let $q = r = 2$, reducing Eq. (6) to

$$t' = t \oplus (c_1c_2 \vee c_3c_4) = t \oplus c_1c_2 \oplus c_3c_4 \oplus c_1c_2c_3c_4, \tag{7}$$

but also

$$t' = t \oplus 1 \oplus \overline{c_1c_2} \cdot \overline{c_3c_4} = t \oplus c_1c_2 \oplus \overline{c_1c_2}c_3c_4. \tag{8}$$

Both expressions of Eq. (7) and Eq. (8) lead to the realizations presented in Fig. 11.

The circuit 12(a) represents a straightforward decomposition. It uses two ancillae and has a quantum cost of 21. The circuit 12(b) is based on the $GF(2)$ equivalence of a disjunction; it does not require ancillae, but has a quantum cost of 39, because of the Toffoli gate with 4 controls. The circuit 12(c) is also based on decomposition, it uses an ancilla, and uses as sub-circuit, the realization of Fig. 7, reaching a total quantum cost of 22. The circuit 12(c) represents the best tradeoff with respect to ancillae and quantum cost.

4 Closing Remark

The concept of hybrid control was introduced, combining conjunctive and disjunctive control of gates. Quantum models of the corresponding gates were disclosed and their complexity analyzed. It is reasonable to expect that hybrid-controlled gates, whenever applicable, will add flexibility to the design of reversible/quantum circuits.

Example Design a circuit for

$$f_1(x_1, x_2, x_3, x_4) = x_1x_2 \oplus \overline{x}_1x_2x_3 \oplus x_1\overline{x}_2x_4 \oplus x_1x_2\overline{x}_3\overline{x}_4, \tag{9}$$

and

$$f_2(x_1, x_2, x_3, x_4, x_5) = x_5 \oplus \overline{x}_5 f_1. \tag{10}$$

Notice that expressing (9) and (10) in positive polarity leads to:

$$\begin{aligned} f_1 &= x_1x_2 \oplus (1 \oplus x_1)x_2x_3 \oplus x_1(1 \oplus x_2)x_4 \oplus x_1x_2(1 \oplus x_3)(1 \oplus x_4) \\ &= x_1x_2 \oplus x_2x_3 \oplus x_1x_2x_3 \oplus x_1x_4 \oplus x_1x_2x_4 \oplus x_1x_2 \oplus x_1x_2x_3 \oplus x_1x_2x_4 \oplus x_1x_2x_3x_4 \\ &= x_2x_3 \oplus x_1x_4 \oplus x_1x_2x_3x_4 \qquad (11) \\ &= x_1x_4 \vee x_2x_3, \end{aligned}$$

$$f_2 = x_5 \oplus (1 \oplus x_5) f_1 = x_5 \oplus f_1 \oplus x_5 f_1 = x_5 \vee f_1. \tag{12}$$

Figure 12 shows realizations based on Eqs. (9) and (10), as well as on Eqs. (11) and (12).

The circuit on the left, which is based on the $GF(2)$ polynomial expressions as given, without optimization, has a quantum cost of 66, since it comprises two Toffoli gates with two controls ($QC = 10$), two Toffoli gates with three controls ($QC = 26$), one Toffoli gate with 4 controls ($QC = 29$), and a $CNOT$ gate ($QC = 1$). The circuit on the right, which is hybrid-controlled, has a quantum cost of 27, since it comprises two Peres gates ($QC = 8$), a Toffoli gate with three controls ($QC = 13$), a $CNOT$ gate ($QC = 1$), and an OR-Toffoli gate ($QC = 5$).

Notice that since the example refers to the generation of two functions, the target lines are initialized with 0. In the hybrid-controlled circuit, up to the inverse Peres gate, the circuit corresponds to the one shown in Fig. 11c, and it uses the line intended to be the target line for f_2 as its ancilla. Since the 0 input is recovered by the inverse Peres gate, the line becomes a correct target line for f_2 and the circuit does not require an additional ancilla.

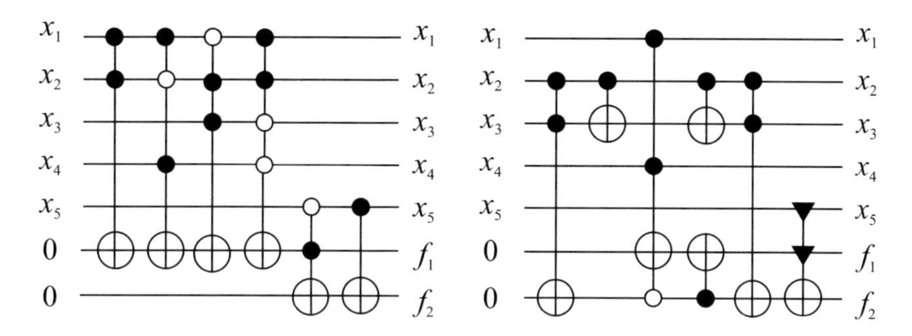

Fig. 12 Circuits for the functions of the example. Left: $GF(2)$ based realization. Right: Hybrid-controlled realization

References

1. Amy, M., Maslov, D., Mosca, M., Roetteler, M.: A meet-in-the-middle algorithm for fast synthesis of depth-optimal quantum circuits. IEEE Trans. Comput. Aided Des. Integr. Circ. Syst. **32**, 818–830 (2013)
2. Barenco, A., Bennett, C.H., Cleve, R., Di Vincenzo, D.P., Margolus, N., Shor, P., Sleator, T., Smolin, J.A., Weinfurter, H.: Elementary gates for quantum computation. Phys. Rev. A **52**, 3457–3467 (1995)
3. Gosset, D., Kliuchnikov, V., Mosca, M., Russo, V.: An algorithm for the T-Count. arXiv quant-ph: 1308.4134v1 (2013)
4. Maslov, D., Dueck, G.W., Miller, D.M., Negrevergne, C.: Quantum circuit simplification and level compaction. IEEE Trans. CAD Integr. Circ. Syst. **27**(3), 436–444 (2008)
5. Moraga, C.: Hybrid GF(2) – boolean expressions for quantum computing circuits. In: De Vos, A., Wille, R. (eds.), RC 2011, LNCS 7165, pp. 54–63. Springer, Berlin (2012)
6. Moraga, C.: Mixed Polarity Reversible Peres gate. Electron. Lett. **50**(14), 987–989 (2014)
7. Peres, A.: Reversible logic and quantum computers. Phys. Rev. A **32**, 3266–3276 (1985)
8. Soeken, M., Thomsen, M.K.: White dots do matter: rewriting reversible logic circuits. In: Dueck, G.W., Miller, D.M. (eds), RC 2013. LNCS 7948, pp. 196–208. Springer, Berlin (2013)
9. Toffoli, T.: Reversible computing. In: Bakker, J.W., van Leeuwen, J. (eds.), ALP 1980. LNCS 84, pp. 632–644. Springer, Berlin (1980)

GenMul: Generating Architecturally Complex Multipliers to Challenge Formal Verification Tools

Alireza Mahzoon, Daniel Große, and Rolf Drechsler

1 Introduction

Nowadays, arithmetic circuits play a key role in many computation intensive applications (e.g., signal processing and cryptography) as well as in upcoming AI architectures (e.g., for machine learning or deep learning). Integer multiplication is one of the most dominant operations in arithmetic circuits, making multipliers crucial components in almost every design. In order to satisfy the demands for fast, area-efficient, and low power designs, a large variety of multiplication algorithms have been introduced. Although some of the proposed algorithms are straightforward, most of them result in highly parallelized and complex hardware architectures. This makes multipliers prone to design errors.

The famous Pentium FDIV bug back in 1994 caused a significant financial loss for the manufacturer. Since then, researchers have put a lot of effort in the development of verification methods in order to avoid design bugs. Particularly, the formal verification methods gained more attention as they take advantage of a mathematical approach to cover the whole input space. Despite the recent achievements, the development of formal verification techniques to support different multiplier architectures is still a big challenge.

In the last 20 years, several verification methods have been proposed to formally prove the correctness of the multipliers. However, most of them suffer from serious limitations: (a) *Decision Diagrams* (DDs) (both bit-level diagrams such as

A. Mahzoon (✉) · R. Drechsler
Institute of Computer Science, University of Bremen, Bremen, Germany
e-mail: mahzoon@informatik.uni-bremen.de; drechsle@informatik.uni-bremen.de

D. Große
Institute for Complex Systems, Johannes Kepler University, Linz, Austria
e-mail: daniel.grosse@jku.at

R. Drechsler, D. Große (eds.), *Recent Findings in Boolean Techniques*,
https://doi.org/10.1007/978-3-030-68071-8_9

BDDs[1] [1] and Word-level Decision Diagrams[2] [5] such as *BMDs) fail due to memory blow-up when it comes to the verification of large multipliers, (b) *Boolean Satisfiability* (SAT) and *Satisfiability Modulo Theories* (SMT) [6] stuck when the bit-width of multiplier increases, (c) *Theorem Proving* [7] needs considerable manual effort before ensuring the correctness, (d) reverse engineering approaches using *Arithmetic Bit-Level* (ABL) [8, 9] are exponential in detection of carry propagation hardware and therefore cannot support a wide range of architectures, and (e) *term rewriting techniques* [10] are not fully automated as a manual update of rewrite rules is necessary for non-existing implementations.

Recently, *Symbolic Computer Algebra* (SCA) approaches have overcome many limitations of the just mentioned methods (see for instance [11–18]). The general idea of SCA-based verification is to

1. Represent the function of the multiplier based on its inputs and outputs as a *Specification Polynomial* SP
2. Capture the logical gates of the circuit also as a set of polynomials P_G
3. Take advantage of Gröbner basis theory in order to prove the membership of SP in the ideal generated by P_G.

The just mentioned third step consists of the step-wise division of SP by P_G known as *backward rewriting*, and eventually the evaluation of the resulting remainder. If this remainder is zero, the multiplier is correct; otherwise, it is buggy.

While SCA-based verification scales for trivial multipliers, non-trivial multipliers, i.e., architecturally complex multipliers, are still problematic for SCA verifiers since for them the number of monomials always explodes during backward rewriting. Recent results showed that this explosion is caused by redundant monomials, known as vanishing monomials in literature [13, 15]. A new theory for the source of vanishing monomials has been introduced in [19]. The vanishing monomials are formed when substituting a converging gate during backward rewriting, i.e., a gate where both outputs of a *Half-Adder* (HA) converge. Reverse engineering techniques to identify atomic blocks allow to further improve this approach such that also dirty multipliers can be verified with up to 1024 output bits [20]. When it comes to the verification of optimized and technology mapped multipliers including

[1] Bryant already proved in his seminal paper introducing BDDs [1] that the Boolean function for the middle output bit of the binary multiplication function has only exponential BDD representations.

[2] Word-level Decision Diagrams represent integer-valued functions $f : \{0, 1\}^n \rightarrow \mathbb{Z}$. Many different types of Word-level Decision Diagrams have been introduced, e.g., MTBDDs, EVBDDs, BMDs, *BMDs, K*BMDs, and *PHDDs [2]. BMDs, *BMDs, K*BMDs, and *PHDDs have the advantage that they provide efficient representations for the multiplier function $mult : \{0, 1\}^{2n} \rightarrow \mathbb{Z}$ mapping two n-bit operand vectors to the number representing the product of the two operands. Although there are papers proposing an efficient construction of Word-level Decision Diagrams from multiplier circuits by a so-called backward construction that starts from a Word-level decision diagram representing the "output word" $\sum_{i=0}^{2n-1} z_i 2^i$ and performs substitutions of gate functions in reverse topological order [3, 4], even intensive efforts could not confirm any practical success of this approach for non-trivial multipliers.

industrial benchmarks, the size of the intermediate polynomial grows very fast, as the boundaries for many atomic blocks have already been destroyed as a result of optimization. Thus, using a dynamic substitution order is mandatory to control the size of the intermediate polynomial [21].

Despite the very good progress, still more insight into multiplier verification in general is mandatory, as we show in several experiments in this chapter. For this purpose, we have developed the multiplier generator GENMUL which can generate very different multiplier architectures based on a wide set of algorithms. This allows the community a broad comparison when verifying different multiplier architectures. Related to GENMUL is the Arithmetic Module Generator, known as AOKI [22]. However, AOKI does only support a maximum input width of 64 bits per multiplier input, and is also closed source. In contrast, the input size of GENMUL is not bounded and we have made GENMUL open source under MIT-license.[3]

This chapter is structured as follows: Sect. 2 reviews the three-stage structure of integer multipliers as well as their formal verification using SCA. Afterwards, Sect. 3 introduces GENMUL. This includes the main data structures available for defining the respective multiplier stage as well as the already supported multiplier architectures. The challenges of verifying different multiplier architectures generated by GENMUL are demonstrated by several conducted experiments and discussed in Sect. 4. Finally, Sect. 5 concludes the chapter.

2 Preliminaries

In this section, we first introduce the general structure of multipliers. Then, we review the basic concepts of SCA. Finally, we explain the SCA-based verification of multipliers in details.

2.1 Multiplier Architectures

Since the invention of the first integer multiplier, the demands for fast and area-efficient designs have encouraged designers to implement a wide variety of multiplier architectures. Although these architectures are apparently very different, they mostly follow the basic idea of computing partial products and then summing the partial products together in order to generate the final result. The main factor differentiating the architectures is the way that they generate and reduce partial products. This process is usually done in three phases which makes the three-stage structure of the multipliers dominant in most designs.

[3]GENMUL is available on http://www.sca-verification.org/genmul.

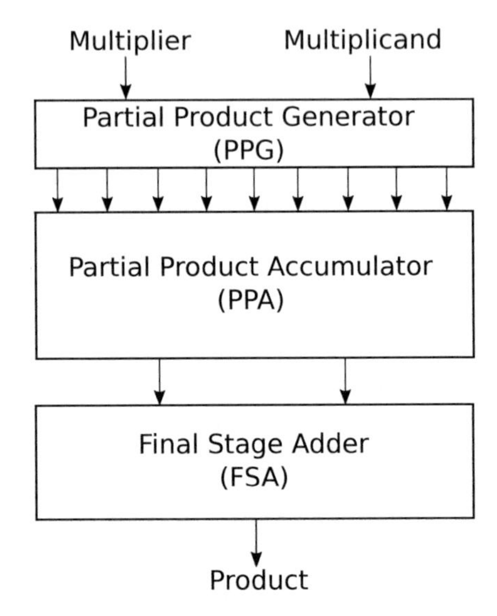

Fig. 1 General three-stage multiplier structure

Figure 1 shows the *three-stage structure* of an integer multiplier. These stages are:

(S1) *Partial Product Generator* (PPG), which generates partial products from the inputs *Multiplier* and *Multiplicand*
(S2) *Partial Product Accumulator* (PPA), which reduces partial products by multi-operand adders and computes their sum
(S3) *Final Stage Adder* (FSA), which converts this sum to the corresponding binary output

There are always some critical parameters in the design of multipliers such as area, delay, and power. These parameters play a major role in determining which architecture/algorithm is suitable for a specific stage [23, 24]. For example, in the second stage, Wallace tree is known for its optimal computational time (lowest overall delay); however, it requires the largest wiring tracks. On the other hand, Balanced delay tree has the smallest number of wiring tracks but suffers from a high overall delay. In the final stage, Ripple carry adder has the most straightforward implementation. In contrast, Carry look-ahead adder enjoys the parallel carry generation which reduces the overall delay of the circuit. Moreover, Lander-Fischer and Brent-Kung adders are recognized for minimum logic depth and minimum area, respectively.

2.2 SCA Basics

Definition A *Monomial* is the power product of variables in the following form:

$$M = x_1^{\alpha_1} x_2^{\alpha_2} \ldots x_n^{\alpha_n} \quad with \quad \alpha_i \in \mathbb{N}_0 \tag{1}$$

A monomial with a coefficient is called a *Term*. □

Definition A *Polynomial* is a finite sum of monomials with coefficients in field k:

$$P = c_1 M_1 + c_2 M_2 + \cdots + c_j M_j \quad c_j \in k \tag{2}$$

A polynomial has a monomial order which facilitates the polynomial manipulations. This order is specified based on the ordering of variables and their powers. We use $A > B$ to show that A is in a higher order than B. For example, in $f = y^4 z + y^2 z^2 + xy$, if we assume that the ordering of the variables is $x > y > z$, then the monomial order will be $xy > y^4 z > y^2 z^2$. The first monomial and the first term after ordering are called *leading monomial* and *leading term* and are denoted by $LM(P)$ and $LT(P)$, respectively.

In SCA, division is denoted by $p \xrightarrow{F} r$, where F is a set of polynomials and r is the remainder. For example, if $p = xy$, $f_1 = x - z$, and $f_2 = yz$, then $xy \xrightarrow{f_1} yz \xrightarrow{f_2} 0$. To perform the division of xy by f_1, first f_1 is multiplied by y to produce the same leading monomial xy as p, so $f_1 y = xy - yz$. Subsequently, the subtraction is performed, i.e., $p - (f_1 y) = xy - (xy - yz) = yz$, which is the result of the first division. Finally, yz is divided by f_2 to get remainder 0.

2.3 SCA-Based Verification

In SCA-based verification of arithmetic circuits, the gate-level netlist and the specification polynomial are given as inputs, and the task is to formally prove that the specification polynomial and the arithmetic circuit are equivalent. The *specification polynomial* is a polynomial determining the function of an arithmetic circuit based on its inputs and outputs. For example, the specification polynomial for the 2-bit multiplier of Fig. 2a is $SP = 8Z_3 + 4Z_2 + 2Z_1 + Z_0 - (2a_1 + a_0)(2b_1 + b_0)$ where $8Z_3 + 4Z_2 + 2Z_1 + Z_0$ describes the 4-bit output, and $(2a_1 + a_0)(2b_1 + b_0)$ indicates the multiplication of the 2-bit inputs.

The gates of an arithmetic circuit can be modeled as polynomials determining the relation between output and inputs. The polynomials of basic Boolean gates are as follows:

$$\begin{aligned} z &= \neg a \Rightarrow p_g := z - 1 + a, \qquad & z = a \vee b \Rightarrow p_g := z - a - b + ab, \\ z &= a \wedge b \Rightarrow p_g := z - ab, \qquad & z = a \oplus b \Rightarrow p_g := z - a - b + 2ab \end{aligned} \tag{3}$$

The polynomials in (3) are in the form of $P_g = x - tail(P_g)$ where x is the gate's output and $tail(P_g)$ is a function based on the gate's inputs.

The gate polynomials for the 2-bit multiplier of Fig. 2a are:

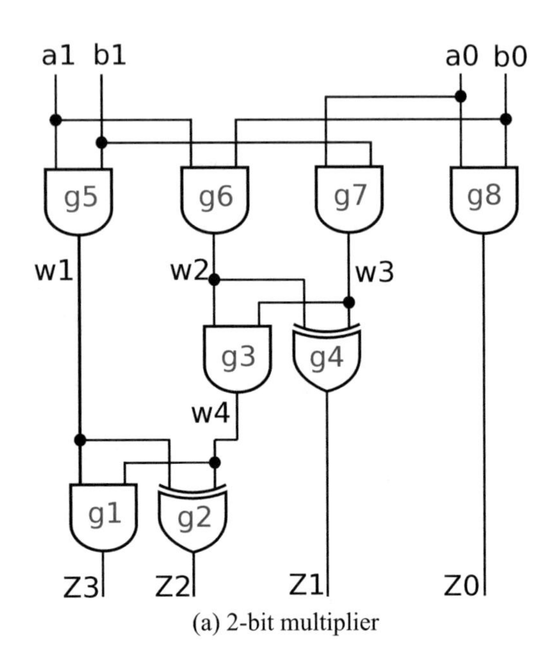

(a) 2-bit multiplier

$$
\begin{aligned}
SP &:= 8Z_3 + 4Z_2 + 2Z_1 + Z_0 - (4a_1b_1 + 2a_1b_0 + 2a_0b_1 + a_0b_0)\\
SP &\xrightarrow{p_{g_1}} SP_1 := 8w_1w_4 + 4Z_2 + 2Z_1 + Z_0 - (4a_1b_1 + 2a_1b_0 + 2a_0b_1 + a_0b_0)\\
SP_1 &\xrightarrow{p_{g_2}} SP_2 := 4w_1 + 4w_4 + 2Z_1 + Z_0 - (4a_1b_1 + 2a_1b_0 + 2a_0b_1 + a_0b_0)\\
SP_2 &\xrightarrow{p_{g_3}} SP_3 := 4w_1 + 4w_2w_3 + 2Z_1 + Z_0 - (4a_1b_1 + 2a_1b_0 + 2a_0b_1 + a_0b_0)\\
SP_3 &\xrightarrow{p_{g_4}} SP_4 := 4w_1 + 2w_2 + 2w_3 + Z_0 - (4a_1b_1 + 2a_1b_0 + 2a_0b_1 + a_0b_0)\\
SP_4 &\xrightarrow{p_{g_5}} SP_5 := 2w_2 + 2w_3 + Z_0 - (2a_1b_0 + 2a_0b_1 + a_0b_0)\\
SP_5 &\xrightarrow{p_{g_6}} SP_6 := 2w_3 + Z_0 - (2a_0b_1 + a_0b_0)\\
SP_6 &\xrightarrow{p_{g_7}} SP_7 := Z_0 - (a_0b_0)\\
SP_7 &\xrightarrow{p_{g_8}} r := 0
\end{aligned}
$$

(b) Backward rewriting steps

Fig. 2 2-bit multiplier and backward rewriting steps. (**a**) 2-bit multiplier. (**b**) Backward rewriting steps

$$
\begin{aligned}
p_{g_1} &:= Z_3 - w_1w_4 & p_{g_5} &:= w_1 - a_1b_1\\
p_{g_2} &:= Z_2 - w_1 - w_4 + 2w_1w_4 & p_{g_6} &:= w_2 - a_1b_0\\
p_{g_3} &:= w_4 - w_2w_3 & p_{g_7} &:= w_3 - a_0b_1\\
p_{g_4} &:= Z_1 - w_2 - w_3 + 2w_2w_3 & p_{g_8} &:= Z_0 - a_0b_0
\end{aligned}
\tag{4}
$$

Assume that the signals of an arithmetic circuit are ordered based on the reverse-topological order (i.e., from outputs toward inputs). The specification polynomial

SP and the gate-level netlist are equivalent, iff the remainder of dividing SP by gate polynomials becomes zero. This division is known as Gröbner basis reduction. For the theory of Gröbner basis and its application to verification of arithmetic circuits, we refer to [15, 25].

The steps of dividing SP by $p_{g_1}, \ldots, p_{g_8}$ for the 2-bit multiplier of Fig. 2a are shown in Fig. 2b. The final remainder of the division is equal to zero, hence the multiplier is bug-free. Please note that all variables in the polynomials are Boolean. Thus, x^n can be replaced by x. Furthermore, for integer arithmetic circuits, dividing SP_i by a gate polynomial $p_{g_i} = x_i - tail(p_{g_i})$ is equivalent to *substituting* x_i with $tail(p_{g_i})$ in SP_i. For example, to obtain the result of the first division step in Fig. 2b, Z_3 can be substituted with $w_1 w_4$ in SP. The process of dividing the specification polynomial by gate polynomials (or equivalently substituting gate polynomials in the specification polynomial) is called *backward rewriting*.

3 Multiplier Generator GENMUL

In this section, we first provide an overview on GENMUL including the main data structures. Then, we describe the supported multiplier architectures that can be generated in the form of Verilog netlists using GENMUL.

3.1 Overview and Data Structures

The main idea behind most of multiplication algorithms is generating and reducing partial products. The partial products are characterized by a *weight*[4] w. Therefore, the integer value of a partial product is $p2^w$, where p is a Boolean number. The partial products with the same weights can be added together to generate the reduced set. In the C++-implementation of GENMUL, we have defined a class named `Partial` containing the two data members `weight` and `ID`, respectively. `ID` is an integer identification number assigned to a partial product automatically during its initialization. `ID` numbers are unique for each partial product, and help us to implement them as wires with specific names in the final Verilog file.

The generation and reduction of partial products are performed by some computational components in each stage of a multiplier. In the first stage, AND gates are the main components. *Half-Adders* (HAs), *Full-Adders* (FAs), and larger adders, e.g., (7:3) counters, are used in the second stage to add the partial products. Finally, multiplexers and carry propagation hardware as well as HAs and FAs construct the final stage of a multiplier. A parent class named `Component` has been defined in GENMUL covering all possible existing computational components. This class has

[4] In literature sometimes also termed significance.

two data members `inputs` and `outputs`, which are vectors of partials and play the role of component inputs and outputs. An existing computational component, e.g., an FA, is implemented as a class inheriting from the parent class `Component`. All these classes contain functions to evaluate outputs based on the partial inputs and the type of the component, and also generate the Verilog code of the component. Besides this, it is possible for the user to add new components to the GenMul and use them for implementing of the architectures.

The availability of the `Partial` and `Component` classes in GenMul allows to easily add further multiplier algorithms to form new multiplier architectures. For example, the Wallace tree takes advantage of a parallel addition algorithm to reduce partial products using HAs and FAs. Hence, the key steps in the implementation are:

1. Adding the partial products with the same weights according to the Wallace algorithm
2. Collecting the outputs of components as new partial products and hashing them based on the weights
3. Repeating the first and second steps until having maximum two partial products with the same weights.

3.2 Generation of Multipliers

GenMul already supports several architectures/algorithms for each stage of a multiplier. We are also working on extra architectures to increase the diversity of the designs even more. Table 1 shows the multiplier architectures of GenMul for three stages of a multiplier. All architectures shown in black are ready to be used. The architectures shown with gray background are under-development, and will be available soon.

One of the main features of GenMul is the generation of multipliers with arbitrary input sizes. Moreover, GenMul is open source on GitHub and due to generic data structures new architectures/algorithms can be easily added. Finally, the web-based version of GenMul is available. We have used the Emscripten toolchain [26] to compile Javascript from our C++ implementation of GenMul. This allows to configure an available multiplier of GenMul on our webpage, then to press the "Generate" button (see Fig. 3) and after generation, the user can directly download the requested Verilog file.

4 Challenges of Verifying Multipliers

In this section, we give insight into challenges of verifying different multiplier architectures. In order to gain a comprehensive understanding, we first generated

Table 1 Multiplier architectures of GENMUL

First stage (PPG)	Second stage (PPA)	Third stage (FSA)
Unsigned simple PPG	Array	Ripple carry
Signed simple PPG	Wallace tree	Carry look-ahead
Unsigned booth PPG	Dadda tree	Lander-Fischer
Signed booth PPG	Counter-based Wallace	Kogge-Stone
	Balanced delay tree	Brent-Kung
	Overturned-stairs tree	Carry-skip
		Han-Carlson
		Carry select
		Conditional sum

24 multipliers with 40 bits per input using GENMUL which cover all possible combinations of different algorithms for each stage.[5] Then, we verified the generated benchmarks by employing eight state-of-the-art formal verification methods. The verification run-times are reported in Table 2. The experiments have been carried out on an Intel(R) Core(TM) i5-4300M CPU 2.60 GHz with 16 GByte of main memory.

The first column of the table shows the **Multiplier architectures**, which consists of three subcolumns: *First stage (PPG)* shows the partial product generator algorithm. *Second stage (PPA)* refers to the employed algorithm for the partial product accumulator. The used algorithm for the final stage adder is shown in the *Third stage (FSA)*.

The run-times (in seconds) of the state-of-the-art verification methods are reported in the second column **Verification methods**, which consists of eight subcolumns: The first seven subcolumns report the run-times of the resent SCA-based verification techniques. The eighth subcolumn *Comm.* refers to the run-time of the commercial formal verification tool OneSpin. Please note that the *Time-Out (TO)* has been set to **10 h**. Moreover, *Failed* in Table 2 implies that verification technique could not complete the task due to an internal error.

We now discuss what can be seen when looking into the results of Table 2. As far as the second and third stages of a multiplier consist of only HAs and FAs, most of the verification techniques successfully verify the multiplier. Multipliers with Array, Wallace, and Dadda algorithms in the second stage and Ripple carry in the third stage are examples of this observation. The reason for this successful verification is: HAs and FAs are two relatively small computational components which have a simple word-level relation between their inputs and outputs:

[5]For this chapter we only used 40 bits per multiplier input, since this already shows the challenges. In [20] we report results for benchmarks generated with GENMUL for up to 512×512 multipliers.

$$\begin{aligned} HA(in : X, Y \quad out : C, S) \quad &\Rightarrow \quad 2C + S = X + Y \\ FA(in : X, Y, Z \quad out : C, S) \quad &\Rightarrow \quad 2C + S = X + Y + Z \end{aligned} \tag{5}$$

Based on (5), substituting the polynomial $2C + S$ of a HA (FA) by its gates polynomials finally results in a polynomial consisting of the inputs summation. Therefore, after substituting/dividing all gates polynomials of an HA (FA), the size of the specification polynomial increases by zero (one). The integration of reverse-engineering in [14, 20] leads to even faster run-times as the HAs and FAs are identified first, and the polynomials of (5) can be used directly during backward rewriting. The column-wise verification methods introduced in [15, 16] work when the multiplier consists of Array and Ripple carry at the second and third stages. However, they time-out if the parallel partial product reduction algorithms such as Wallace or Dadda tree are used for the second stage.

Despite Ripple carry adder, some of the algorithms used in the third stage of the multiplier take advantage of the carry propagation hardware to reduce the overall delay. Carry look-ahead and parallel prefix adders (Lander-Fischer, Kogge-Stone, and Brent-Kung) are among these architectures. Based on our experiments, a huge number of vanishing monomials are generated during verification of multipliers containing carry propagation hardware which results in polynomial explosion during backward rewriting. To illustrate the role of carry propagation hardware in generation of vanishing monomials, we consider the Boolean formulation of a 4-bit carry look-ahead adder:

$$\begin{aligned} G_i &= x_i \wedge y_i, \\ P_i &= x_i \oplus y_i, \\ c_1 &= G_0 \vee (c_0 \wedge P_0), \\ c_2 &= G_1 \vee (G_0 \wedge P_1) \vee (c_0 \wedge P_0 \wedge P_1), \\ c_3 &= G_2 \vee (G_1 \wedge P_2) \vee (G_0 \wedge P_1 \wedge P_2) \vee (c_0 \wedge P_0 \wedge P_1 \wedge P_2), \\ c_4 &= G_3 \vee (G_2 \wedge P_3) \vee (G_1 \wedge P_2 \wedge P_3) \vee (G_0 \wedge P_1 \wedge P_2 \wedge P_3) \\ &\quad \vee (c_0 \wedge P_0 \wedge P_1 \wedge P_2 \wedge P_3) \end{aligned} \tag{6}$$

where x_i and y_i are ith bit of the first and second input and c_i is the final carry.

In (6), G_i and P_i are outputs of an HA. Hence, the product of them is always equal to zero, i.e., $G_i P_i = 0$. If c_4 is transformed naively to polynomial form, it consists of 31 monomials. However, 26 monomials contain the product of G_i and P_i, i.e., they reduce to zero after G_i and P_i substitution. These monomials are vanishing monomials and cause explosion in the number of monomials before cancellation. The proposed methods of [11, 14–16] do not provide any solution for this vanishing monomial problem. As a result, they time-out in verification of multiplier architectures containing carry look-ahead, Lander-Fischer, Kogge-Stone, and Brent-Kung (see Table 2). The XOR-rewriting heuristic in [13] enables us to verify some of these architectures. However, this method is not robust and fails for many benchmarks since it misses many vanishing monomials. The idea of creating converging gate cones and removing vanishing monomials locally before global backward rewriting as introduced in [19] helps to remove all vanishing monomials and verify most of the non-trivial architectures. Finally, the proposed method of [20], which integrates reverse engineering and local vanishing removal, supports the verification of more architectures and is much faster than [19] as can be seen in Table 2.

All SCA-verification methods fail to verify multipliers using counter-based Wallace algorithm in the second stage. Counters are computational components that count the number of inputs whose value is 1 and return the result on outputs; e.g., HA and FA are (3:2) and (2:2) counters, respectively. Counter-based Wallace algorithm takes advantage of bigger counters (usually (7:3) counters) to speed up the partial product reduction. Counters still have a simple relation between inputs and outputs. However, they constitute of a larger number of logical gates in comparison to HAs and FAs. Therefore, the large number of generated monomials during substitution of intermediate gates polynomials leads to the verification failure. We plan to extend the reverse engineering technique of RevSCA [20] in order to identify counters with different sizes. Finally, please note that the commercial tool could not verify any of the generated benchmarks.

Overall, as can be concluded from the experiments, additional research is needed to conquer the automatic formal verification of multipliers.

5 Conclusion

In this chapter, we have introduced the multiplier generator GenMul which allows generation of a wide range of multiplier architectures. GenMul is open source and already supports many different algorithms for each stage of multipliers. Moreover, due to its generic data structures, further algorithms can be easily added. We used the generated multipliers of GenMul to challenge all available recent SCA-based formal verification techniques, and also provided some insight on the reasons why certain approaches fail in verifying non-trivial multiplier architectures.

Acknowledgements This work was supported by the German Research Foundation (DFG) within the project VerA (GR 3104/6-1 and DR 297/37-1).

Table 2 Verification run-times for 40-bit multipliers (run-times in seconds)

Multiplier architectures			Verification methods							
First stage (PPG)	Second stage (PPA)	Third stage (FSA)	[20]	[19]	[13]	[11]	[14]	[15]	[16]	Comm.
Simple PPG	Array	Ripple carry	4.73	16.96	47.12	5.88	0.04	165.40	2.36	TO
		Carry look-ahead	12.36	36.88	91.60	TO	TO	TO	TO	TO
		Lander-Fischer	4.90	17.47	50.90	TO	TO	TO	TO	TO
		Kogge-Stone	9.49	TO	Failed	TO	TO	TO	TO	TO
		Brent-Kung	5.01	12.89	Failed	TO	TO	TO	TO	TO
		Carry-skip	14.32	TO	Failed	TO	TO	TO	Failed	TO
	Wallace	Ripple carry	5.56	19.49	Failed	7.17	0.04	TO	TO	TO
		Carry look-ahead	94.71	251.31	Failed	TO	TO	TO	TO	TO
		Lander-Fischer	6.32	19.78	Failed	TO	TO	TO	TO	TO
		Kogge-Stone	41.33	TO	Failed	TO	TO	TO	TO	TO
		Brent-Kung	7.69	18.62	Failed	TO	TO	TO	TO	TO
		Carry-skip	246.79	TO	Failed	TO	TO	TO	TO	TO
	Dadda	Ripple carry	5.91	17.63	51.64	5.59	0.04	TO	TO	TO
		Carry look-ahead	134.29	296.72	TO	TO	TO	TO	TO	TO
		Lander-Fischer	7.65	18.18	60.77	TO	TO	TO	TO	TO
		Kogge-Stone	40.01	TO	Failed	TO	TO	TO	TO	TO
		Brent-Kung	7.48	16.69	58.49	TO	TO	TO	TO	TO
		Carry-skip	228.92	TO	Failed	TO	TO	TO	TO	TO
	Counter-based Wallace	Ripple carry	TO	TO	TO	TO	TO	TO	TO	TO
		Carry look-ahead	TO	TO	TO	TO	TO	TO	TO	TO
		Lander-Fischer	TO	TO	TO	TO	TO	TO	TO	TO
		Kogge-Stone	TO	TO	TO	TO	TO	TO	TO	TO
		Brent-Kung	TO	TO	TO	TO	TO	TO	TO	TO
		Carry-skip	TO	TO	TO	TO	TO	TO	TO	TO

TO: Time-Out of 10 h Failed: Internal error

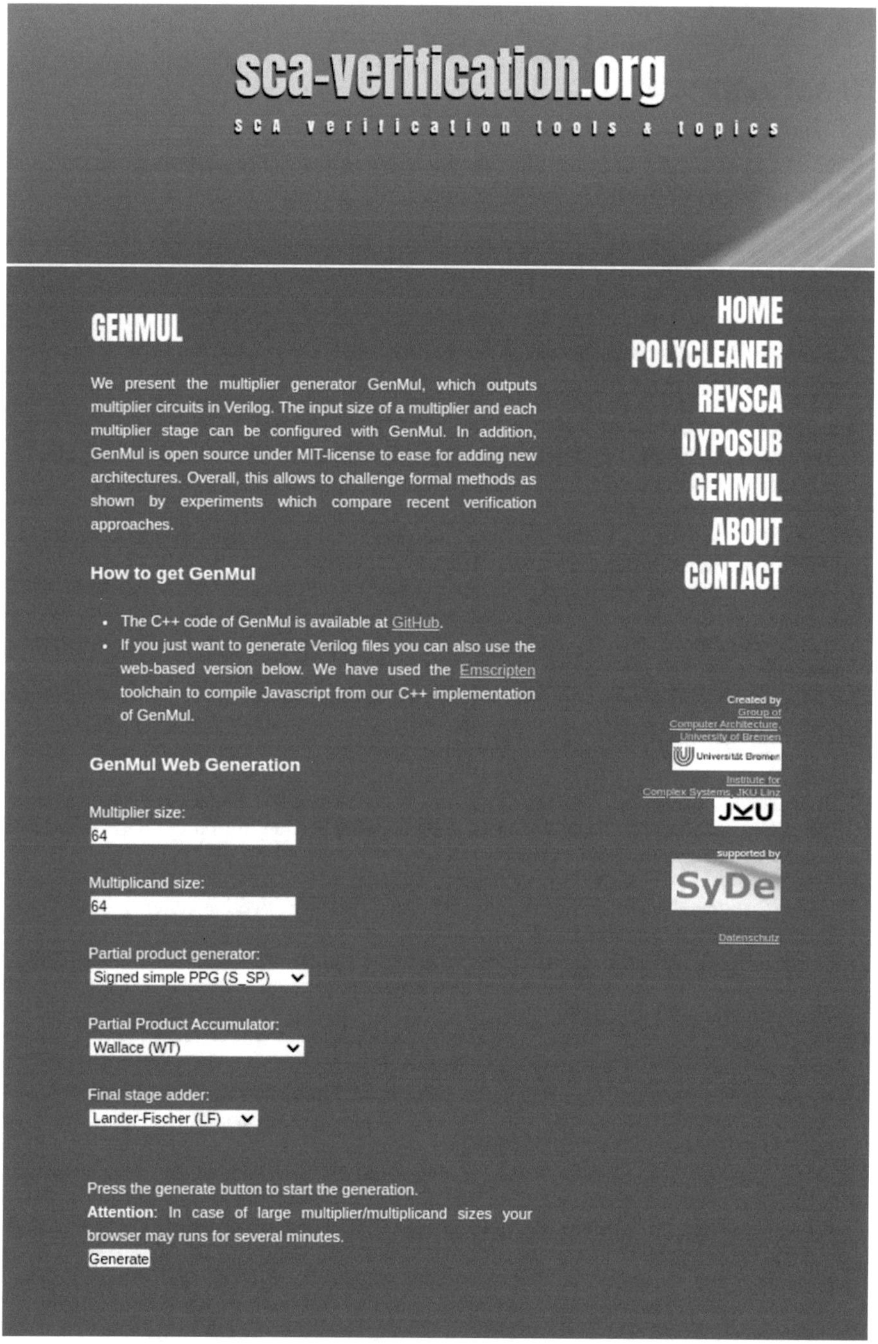

sca-verification.org

SCA verification tools & topics

HOME
POLYCLEANER
REVSCA
DYPOSUB
GENMUL
ABOUT
CONTACT

GENMUL

We present the multiplier generator GenMul, which outputs multiplier circuits in Verilog. The input size of a multiplier and each multiplier stage can be configured with GenMul. In addition, GenMul is open source under MIT-license to ease for adding new architectures. Overall, this allows to challenge formal methods as shown by experiments which compare recent verification approaches.

How to get GenMul

- The C++ code of GenMul is available at GitHub.
- If you just want to generate Verilog files you can also use the web-based version below. We have used the Emscripten toolchain to compile Javascript from our C++ implementation of GenMul.

GenMul Web Generation

Multiplier size:
64

Multiplicand size:
64

Partial product generator:
Signed simple PPG (S_SP)

Partial Product Accumulator:
Wallace (WT)

Final stage adder:
Lander-Fischer (LF)

Press the generate button to start the generation.
Attention: In case of large multiplier/multiplicand sizes your browser may runs for several minutes.
Generate

Created by
Group of
Computer Architecture,
University of Bremen
Universität Bremen
Institute for
Complex Systems, JKU Linz
JKU
supported by
SyDe
Datenschutz

Fig. 3 GENMUL website

Appendix

GenMul *Website*

GenMul is now available on http://www.sca-verification.org/genmul. In Fig. 3, a screenshot of the GenMul website is shown.

References

1. Bryant, R.E.: Graph-based algorithms for Boolean function manipulation. IEEE Trans. Comput. **35**(8), 677–691 (1986)
2. Scholl, C., Becker, B., Weis, T.: On WLCDs and the complexity of word-level decision diagrams – a lower bound for division. Formal Methods Syst. Des. Int. J. **20**(3), 311 (2002)
3. Hamaguchi, K., Morita, A., Yajima, S.: Efficient construction of binary moment diagrams for verifying arithmetic circuits. In: International Conference on Computer-Aided Design, pp. 78–82 (1995)
4. Keim, M., Drechsler, R., Becker, B., Martin, M., Molitor, P.: Polynomial formal verification of multipliers. Formal Methods Syst. Des. **22**(1), 39–58 (2003)
5. Drechsler, R.: Formal Verification of Circuits. Kluwer Academic Publishers, New York (2000)
6. Diao, Y., Wei, X., Lam, T., Wu, Y.: Coupling reverse engineering and SAT to tackle np-complete arithmetic circuitry verification in $\sim$o(# of gates). In: ASP Design Automation Conference, pp. 139–146 (2016)
7. Kapur, D., Subramaniam, M.: Mechanically verifying a family of multiplier circuits. In: Computer Aided Verification, pp. 135–146 (1996)
8. Stoffel, D., Kunz, W.: Equivalence checking of arithmetic circuits on the arithmetic bit level. IEEE Trans. Comput. Aided Des. Circ. Syst. **23**(5), 586–597 (2004)
9. Pavlenko, E., Wedler, M., Stoffel, D., Kunz, W., Wienand, O., Karibaev, E.: Modeling of custom-designed arithmetic components in ABL normalization. In: Forum on Specification and Design Languages (2008), pp. 124–129
10. Vasudevan, S., Viswanath, V., Sumners, R.W., Abraham, J.A.: Automatic verification of arithmetic circuits in RTL using stepwise refinement of term rewriting systems. IEEE Trans. Comput. **56**(10), 1401–1414 (2007)
11. Farahmandi, F., Alizadeh, B.: Gröbner basis based formal verification of large arithmetic circuits using gaussian elimination and cone-based polynomial extraction. Microprocessors Microsyst. **39**(2), 83–96 (2015)
12. Yu, C., Brown, Liu, W.D., Rossi, A., Ciesielski, M.: Formal verification of arithmetic circuits by function extraction. IEEE Trans. Comput. Aided Des. Circ. Syst. **35**(12), 2131–2142 (2016)
13. Sayed-Ahmed, A., Große, D., Kühne, U., Soeken, M., Drechsler, R.: Formal verification of integer multipliers by combining Gröbner basis with logic reduction. In: Design, Automation and Test in Europe, pp. 1048–1053 (2016)
14. Yu, C., Ciesielski, M., Mishchenko, A.: Fast algebraic rewriting based on and-inverter graphs. IEEE Trans. Comput. Aided Des. Circ. Syst. **37**(9), 1907–1911 (2017)
15. Ritirc, D., Biere, A., Kauers, M.: Column-wise verification of multipliers using computer algebra. In: International Conference on Formal Methods in CAD, pp. 23–30 (2017)
16. Ritirc, D., Biere, A., Kauers, M.: Improving and extending the algebraic approach for verifying gate-level multipliers. In: Design, Automation and Test in Europe, pp. 1556–1561 (2018)
17. Mahzoon, A., Große, D., Drechsler, R.: Combining symbolic computer algebra and boolean satisfiability for automatic debugging and fixing of complex multipliers. In: IEEE Annual Symposium on VLSI, pp. 351–356 (2018)

18. Kaufmann, D., Biere, A., Kauers, M.: Verifying large multipliers by combining SAT and computer algebra. In: International Conference on Formal Methods in CAD, pp. 28–36 (2019)
19. Mahzoon, A., Große, D., Drechsler, R.: PolyCleaner: clean your polynomials before backward rewriting to verify million-gate multipliers. In: International Conference on Computer-Aided Design, pp. 129:1–129:8 (2018)
20. Mahzoon, A., Große, D., Drechsler, R.: RevSCA: Using reverse engineering to bring light into backward rewriting for big and dirty multipliers. In: Design Automation Conference, pp. 185:1–185:6 (2019)
21. Mahzoon, A., Große, D., Scholl, C., Drechsler, R.: Towards formal verification of optimized and industrial multipliers. In: Design, Automation and Test in Europe, pp. 544–549 (2020)
22. Arithmetic module generator based on ACG. Available at https://www.ecsis.riec.tohoku.ac.jp/topics/amg/i-amg (2019)
23. Zimmermann, R.: Binary adder architectures for cell-based vlsi and their synthesis. Ph.D. dissertation, Swiss Federal Institute of Technology (1997)
24. Koren, I.: Computer Arithmetic Algorithms, 2nd ed. A. K. Peters, Natick (2001)
25. Cox, D.A., Little, J., O'Shea, D.: Ideals Varieties and Algorithms. Springer, New York (1997)
26. "Emscripten," Available at https://emscripten.org (2019)

Index

A
ABC/BOOLECTOR, 16–17, 21–22
Adder substitution, 12, 18
Affine functions, 105, 106
Algebraic proof system, 8–9, 90
Algorithm
  bidirectional, 151
  Buchberger's, 6
  FFT-like permutation, 105–123
  instance properties, 137
  multi-directional, 151
  NN and genetic, 32
  optimization, 31, 47, 52
  $p_{\min}$ and $n_{\max}$, 75–76
  search, 151–153
AMULET, 10–12, 18–24
Arithmetic circuit, 1, 2, 10, 177, 181–183
ARITHMETIC MODULE GENERATOR, 14, 18–19
Asymptotic theory, 98
Automated reasoning, 3
Axiomatization
  Boolean
    derivatives, 83
    differentiation, 84–85
  infinite models, 84
  model-theoretic
    advantage, 84
    fundamentals, 85–89
  outline, 85

B
Basic transformation-based synthesis method, 150, 162
Benchmarks
  generators
    ABC/BOOLECTOR, 16–17
    ARITHMETIC MODULE GENERATOR, 14
    EPFL Combinational Benchmark Suite, 16
    GenMul, 15
    MULTGEN, 15
    optimizing benchmarks, 17–18
    processing verilog benchmarks, 17
  state-of-the-art formal verification methods, 185
  web site, 161
Bent functions
  algorithm, 122–123
  Boolean functions, 105
  examples, 112–122
  FFT, 110–111
  functional expressions, 105
  non-zero Reed-Muller coefficients, 106
  permutation matrices, 111–112
  spectral
    domain, 106
    invariant operations, 108–109
  Walsh
    coefficients, 106–107
    domain, 108–109
    transform, 107–110
Bentness, 106, 111, 122, 123
Bidirectional transformation-based synthesis method, 151
Bijective transformation, 145
Black triangles, 167
Boolean derivative, 83, 91, 94, 97, 101

R. Drechsler, D. Große (eds.), *Recent Findings in Boolean Techniques*,
https://doi.org/10.1007/978-3-030-68071-8

Boolean differentiation
  applications and perspectives, 101–103
  Boolean functions, rings, and derivations, 89–90
  complete axiomatization, 91–97
  finite models, 97–101
Boolean ring, 87, 89, 90, 93, 94, 96–99, 101, 102
BSIM, 29, 30, 33, 36, 43

C
Carry propagation hardware, 183, 186
Categoricity, 86, 93, 95, 96
Circuit simulation, 29, 31, 33, 41
Circuit verification
  algebra, 5–6
  algebraic proof systems, 8–9
  multiplier circuits, 3–5
  using computer algebra, 7–8
Clique cover, 127–134, 136
Compact model, 29, 30, 32–34, 38, 39, 43, 48, 55
Compatibility graph, 129–132, 134
Complete axioms, 84, 91–97
Computer algebra, 2, 3
  algebra, 5–6
  algebraic proof systems, 8–9
  circuit verification, 7–8
  multiplier circuits, 3–5
Cooley-Tukey FFT, 106, 110, 123
Counter, 8, 14, 15, 183, 185, 187, 188
Cube compatibility, 128, 130
Cube intersection, 128

D
Deep analysis, 65
Design Technology Co-Optimization (DTCO), 31–33, 44, 45–48, 50
  NC-FinFET, 47
  parameter optimization, 50–54
Differential operator, 101, 102
Disjoint translation, 113
Disjunctive control, 168, 169
DTCO, *see* Design Technology Co-Optimization (DTCO)
Dyadic group, 107, 110
Dynamic substitution, 179
DYPOSUB, 13–14, 18–22, 24

E
Elementary symmetric function, 62, 64, 65, 80
Emerging technologies, 30, 32–34, 38
EPFL Combinational Benchmark Suite, 16, 21
Equivalence class, 132
Exact solution, 65, 136
Experimental evaluation, 10, 11, 14, 141

F
F_2-vector space, 87, 88, 92, 95, 99
Factor matrices, 107, 110, 112, 123
  sparse, 111
FA, *see* Full-adder (FA)
Fast Fourier transform (FFT)
  essence of, 110–111
  Good-Thomas factorization, 106
  permutation matrices, 123
  Walsh transform, 106
Fast Walsh transform (FWT), 106, 107, 110, 113, 123
FFT, *see* Fast Fourier transform (FFT)
Final stage adder (FSA), 5, 11, 14, 15, 180, 185
FinFET transistor, 34
First-order theory, 83, 85, 86, 88, 93, 96, 97, 101
Formal verification
  bit-level reverse engineering, 2
  circuit verification, 3–9
  digital circuits, 1
  multipliers, 187
  Pentium FDIV bug, 1
  predefined specification, 2
  processors, 1
  tools
    Algebraic RewriTing in ABC, 10–11
  using SCA, 179
Fraisse limit, 88– 100
FSA, *see* Final stage adder (FSA)
Full-adder (FA), 2, 5, 10, 13, 17, 21, 183
Function extraction, 10
Function inverse, 147–149, 157, 158, 162, 163
Function translations, 143, 144, 159, 160, 163
  discussion and future work, 163–164
  function inverse, 147–148
  heuristic selection, 161–163
  input-output
    negation, 148–149
    permutation, 149

G
Gate constraints, 7, 10
Generator
  ABC/BOOLECTOR, 16–17
  ARITHMETIC MODULE, 14

EPFL Combinational Benchmark, 16
GenMul, 15
MULTGEN, 15
multiplier, 183–184
optimizing benchmarks, 17–18
processing verilog benchmarks, 17
Generic theory, 87, 88, 98, 101
GenMul, 15, 17, 19–20
arithmetic circuits, 177
multiplier
architectures, 179–180
generator, 183–184
verification, 179
Pentium FDIV bug, 177
SCA basics, 180–181
SCA-based verification, 177–178, 181–183
Good-Thomas factorisation, 106, 110, 123
Gröbner basis, 2, 6–8, 10, 11, 178, 183

H
Hadamard order, 106, 107, 109, 110, 123
Half-adder (HA), 5, 12, 13, 178, 183
Hamming weight, 105, 106
HA, *see* Half-adder (HA)
Hybrid control
gates, 169–171
quantum models, 174–175
scalability, 171–173
symbol and functionality, 167–169

I
Ideal membership testing, 6, 7
Index generation function
best method, 60, 61
branch-and-bound approach, 59
computation time, 60
experimental results, 76–80
heuristic approach, 60
linear decomposition, 59
linearity, orthogonality, and circuit structures, 63–64
$p_{\min}$ and $n_{\max}$, 75–76
preliminaries, 61–63
reverse task
analysis, 74–75
missing values, 71–74
regions of restrictions, 68–69
repeated use of the smallest optimal circuits, 70–71
smallest optimal circuits, 66–68
trivial solution for $t = 1$, 66
task to solve, 65
used approach, 65
Infinite Boolean Algebras, 84, 85, 87
Input-output negation, 148–149, 157, 160–163
Input-output permutation, 149, 157, 160–163
Inverse Peres gates, 144, 146, 148, 158, 163, 169, 173, 174
Involution, 84, 90, 91, 93–97, 102, 111

L
Library characterization
ML estimators training, 46–47
training data generation, 45–46
Linearly independent derivatives, 93, 98

M
Machine learning
circuit simulations, 29
compact models, 33–34
design technology co-optimization, NC-FinFET, 47
early evaluation of technology, 30
evaluation and experimental results
cell library prediction accuracy, 48, 49
DTCO parameter optimization, 50–54
performance improvement, 54–55
prediction on system level accuracy, 50
for library characterization, 45–47
foundry secrecy, 29
innovation, 29–30
negative capacitance FinFET, 34
related work, 32–33
standard cell model, 31–32
transistor
characteristics, 34–35
model, 30, 35–44
Mixed controlled gates, 146, 163
Mixed polarity, 145, 146, 159, 163
Mixed-polarity multiple-control Toffoli gates (MPMCT), 145, 159
Mixed-polarity Peres (MPP) gates, 146
Model theory, 83–87, 98
Monomial
backward rewriting, 178
coefficient, 181
modular reasoning, 8
specification, 12
verification
of multipliers, 186
tools, 3
MPMCT, *see* Mixed-polarity multiple-control Toffoli gates (MPMCT)
MULTGEN, 15, 17, 20

Multiplier
  architectures, 179–180
  challenges, 184–186
  circuits (*see* Formal verification)
  design errors, 177
  generator (*see* Multiplier generator)
  GenMul (*see* GenMul)
  SMT-solver BOOLECTOR, 16
Multiplier generator
  challenges, 184–188
  data structures, 183–184
  generation, 184
  overview, 183–184
  website, 187, 189
Multiplier verification, 8, 179
Multi-valued (MV)
  function, 127, 133
  variable, 127, 133

N
NC-FinFET, 30, 34, 36, 38, 43–45, 47, 48, 51, 55
NCV quantum gates, 144, 146, 153, 155
Negative capacitance transistor, 34
Neural Networks (NNs), 30, 39, 42, 46
  and genetic algorithms, 32
  interpolation, 38
  modeling, 30
  NC-FinFET, 43–44
  training data, 36
  transistor modeling, 38–40
Nonlinear codes
  ATE channels, 126
  BIST domain, 126–127
  compatibility graph properties, 129–132
  digital device, 125
  error-correcting code, 126
  expander
    input, 132–134
    outputs as a clique cover problem, 127–129
  FSM, 125
  method summary, 134, 135
  minimum clique cover techniques, 132
  MV-encoding problem, 132–134
  results
    algorithms 1 and 2, 136
    expander synthesis, 138–141
    expander widths comparison, 136, 137
    implementation, 135
    instance properties, 136, 137
    MV encoding, 136, 138
Nullstellensatz proof, 8, 9

O
Open-source, 14, 179, 184, 187
Optimizing benchmarks, 17–18, 22–23
OR-Peres gates, 169
OR-Toffoli gates, 174

P
Partial product, 4, 5, 179, 180, 183–187
Partial product accumulator (PPA), 5, 180, 185, 188
Partial product generator (PPG), 4, 5, 180, 185, 188
Permutation matrices, 107, 109–114, 121–123
  FFT-like, 22, 111, 112, 122
Polarization of function, 108, 110
Polarization of variables, 108, 110, 112
POLYCLEANER, 10, 12, 18–21
Polynomial
  by-product, 3
  calculus, 8
  input and output variables, 7
  intermediate, 179
  internal gate, 12
  monomial order, 181
  Nullstellensatz proof system, 9
  principal ideal domains, 8
  specification, 2
PPA, *see* Partial product accumulator (PPA)
PPG, *see* Partial product generator (PPG)
Proof checking, 3, 9, 24
Proof generation, 23, 24

Q
Quantifier elimination, 89, 99–101, 103
Quantum circuit cost
  MPP and MPIP gate cost and substitution, 156
  NCV realization of MPMCT gates, 155, 156
  negative control CNOTs, 154–155
  simplification and mapping strategy, 156
Quantum cost, 151, 153, 155, 157–159, 161–164, 169, 171, 173, 174
Quantum gates, 144, 146–147, 154, 164, 170
Quantum models, 169, 173, 174
Quantum V-gates, 170
Quantum W-gates, 170

R
Reed-Muller coefficients, 106
Reed-Muller expressions, 105

Reverse engineering, 2, 13, 29, 39, 178, 187
Reverse task, 60, 65–66, 76, 78, 80
Reversible circuit synthesis
  Boolean function, 143
  experimental results, 157–161
  function translations, 143, 147–149
  quantum circuit, 154–156
  quantum gates and circuits, 146–147
  reversible functions, gates, and circuits, 145–146
  simplifying, 153–154
  transformation-based synthesis, 144, 150–153
Reversible circuits
  for function, 146
  mapping, 154–156
  simplifying, 153–154
  synthesis (*see* Reversible circuit synthesis)
Reversible functions, 144–150, 157–160, 162
Reversible gates, 143, 145, 146, 148, 154, 169
REVSCA/REVSCA-2.0, 10, 12–13, 18
Ridge Regressors, 46

S
Scan chain, 125–127
Sequential Least Squares Programming (SLSQP), 47
Smoothly approximable structures, 103
Spectral invariant operations, 106–114, 118, 122, 123
SPICE, 29, 31–33, 35–40, 42
Stability, 38, 84, 86, 87, 103
Standard cells
  library, 18, 22, 31, 33
Survey
  formal verification (*see* Formal verification)
  integer multipliers, 3
Switching algebras, 98–101
Symmetric function, 62, 64, 65, 80

T
Test compression, 126
Test delivery, 125
Test vector redundancy, 126
Three-stage structure, 179, 180
Time complexity, 4, 80
Toffoli and Peres gates, 145, 146, 150, 155, 159, 167–169, 171, 173–175
Tools
  EDA, 30, 31, 44, 47
  formal verification (*see* Formal verification)
  model-theoretic, 93
  verification, 10–14
Transformation-based synthesis, 144, 150–153, 160–163
Transistor modeling
  data scaling, 36–38
  early evaluation with limited data, 42–43
  experimental setup, 35–36
  inference accuracy, 40–41
  machine learning, 30
  NC-FinFET, 43–44
  NN-based
    advantages, 38–39
    disadvantages, 39–40
  traditional fitness, 41–42
  training time, 40–41

V
Vanishing monomials, 12, 178, 186, 187
Variable encoding, 127, 141
Vectorial derivative, 84, 90, 95
Verilog benchmarks, 17
Very fast algorithm, 179

W
Walsh spectrum, 107, 109, 110, 113, 115, 123
  flat, 109
Walsh transform, 106–110, 123
White dots, 144, 167

Printed in the United States
by Baker & Taylor Publisher Services